ENERGY

FROM NATURE
TO MAN

ENERGY

FROM NATURE
TO MAN

William C. Reynolds

Professor and Chairman
Department of Mechanical Engineering
Stanford University

McGraw-Hill Book Company

New York   St. Louis   San Francisco   Düsseldorf   Johannesburg
Kuala Lumpur   London   Mexico   Montreal   New Delhi
Panama   Paris   São Paulo   Singapore   Sydney   Tokyo   Toronto

Library of Congress Cataloging in Publication Data

Reynolds, William Craig, date
  Energy: From Nature to Man

  Bibliography: p.
  1. Power (Mechanics) I. Title
TJ153.R43   621   73-22419
ISBN 0-07-052043-7

ENERGY: From Nature to Man

1234567890KPKP7987654

This book was set in Baskerville.
The editors were B. J. Clark and Matthew Cahill;
the designer was Pencils Portfolio, Inc.;
the production supervisor was Leroy A. Young.
The drawings were done by John Cordes,
J & R Technical Services, Inc.
Kingsport Press, Inc., was printer and binder.

CONTENTS

PREFACE

This book was developed for use in a course about energy and energy technology aimed primarily at nonengineering students. The approach is quantitative but at a modest mathematical level; some high school science and mathematics seems adequate preparation.

At Stanford the course is part of an integrated program called "Values, Technology, and Society" (VTS). The VTS program has several objectives, including the goal of bringing technical literacy to a wide segment of the general undergraduate population. This book is used in conjunction with other readings on energy technology, energy resources, and energy policy and involves contributions from faculty members of several of the engineering departments.

In these pages we have attempted to provide a basic background in the concepts that underlie engineering science and to present these in a manner that students without much technical training can easily relate to personal experience. The objective is *not* to make students skilled at engineering analysis but to give them some feeling and appreciation for what such analysis involves, so that they may better understand what they read. A style of problems has been developed which seems to be quite successful with such students. We ask them to do some simple quantitative analysis, related to some sort of energy problem, and then to reflect upon the significance of their analysis to their society. Sometimes we ask them to take the position of staff analysts for governmental agencies, other times to think as a responsible citizen about to cast a vote or engage in public debate on some critical issue. The objectives are to help them realize the value of good quantitative information in dealing with these situations and to help them develop their own system of values.

We have found that, after developing a modest technical background, these students can read and digest material presented in such semitechnical magazines as *Scientific American* or such trade magazines as *Mechanical Engineering, Power,* and *Nuclear News.* Each student is asked to select an energy technology of his own choice and to prepare a survey article on that subject in a style appropriate for communication to persons with minimal technical backgrounds. Rather remarkable papers on such diverse topics as "cooling towers," "breeder reactors," "refuse-energy systems," "geothermal power," and "electric automobiles" have emerged from this effort. Once acquainted with the literature available in a good engineering library, the students find their own interests and dig remarkably deeply into their chosen subjects.

Many of these students are very inexperienced in their ability to conceptualize hardware. We have found that class time spent visiting our mechanical engineering laboratory, where a vast array of energy hardware is on display and where we can show them ongoing research work in energy sciences, is time very well spent. We also visit our central heating and cooling plant, where students can hear the big boilers, see the steam turbine that generates power for the computer center, feel the warm water falling out of the cooling towers, feel the moist air blown in their faces from the top of the cooling towers, and see the complex system of controls. We also visit a modern oil refinery and a nuclear power station, both of which provide marvelous eye-opening experiences. We have stood in the 450-ft-high cooling towers of a nuclear power plant under construction, been inside the reactor vessel, and seen the steel and concrete and all the machinery required to pour energy into three tiny wires that trail out from the plant. These experiences are invaluable to students thirsty for an understanding of energy technology, and instructors using this material are strongly urged to arrange similar experiences for their students.

If you are an instructor about to start a course like this, always guard against being overly superficial; this is a very troublesome point, but one that need not be a problem if you ask each student to develop some depth in a particular narrow area. The term papers on specific technologies are very helpful in this regard. Another useful approach is to bring in persons with expertise in fields other than your own. You may know the anecdotes in your field of energy, but probably not in others; your guests will field questions better, can supplement the text material, and can help bring depth to the subject in areas where you, yourself, just aren't sufficiently experienced. But your organization and coordination of your visitors will be essential to success in your course; be sure that your visitors know precisely where your students will be at the time they come, and if possible go over their talks with them in detail ahead of time to eliminate unnecessary duplication and to help identify the points that need to be emphasized most strongly.

If you are a student about to study this book on your own, don't be content with the information contained in these pages; it is just your starter. Get to a library and read some of the magazines mentioned; talk to engineers about their work. Do some of the personal experiments given as problems. Only by such a concerted effort will you develop a really useful knowledge and perspective of the very interesting and complex technology that brings energy from nature to man.

William C. Reynolds

ENERGY

FROM NATURE
TO MAN

I

THE BIG PICTURE

In which we get
some idea where things
and we are going

OBJECTIVES, SEMANTICS, PERQUISITES

This is a book about energy. Between 1960 and 1970 we in the United
States consumed as much electrical energy as in all time before 1960.
It is quite likely that between 1970 and 1980 we again will consume more
electrical energy than in all time before 1970. We cannot do this without
impact on ourselves, our environment, and the world. Drastic changes
in the trends of energy utilization must somehow be made in the years
not far ahead. These matters are already becoming the subject of intense
public debate. To act responsibly, and intelligently, one must develop
an understanding of what energy is, how we use it, and its impact upon
society; only then can one participate rationally in the decisions that
must be made.

Rational thinking about a subject such as energy has two important aspects. First, there are *quantitative matters* for which numbers can be developed by computation based on scientific principles. One might think that there is not much room for argument here; but there is, for to make such a calculation one must always approximate the real world by a simplified model, and then use data (often incomplete) gathered from measurements (often inaccurate). A well-trained engineer or scientist will be able to identify the pitfalls of a particular quantitative argument, and will produce a number of analyses covering the range of probable realities. Then there is room for considerable debate as to which end of the range is most realistic, and the debate can get very heated when one end spells doom and the other paradise. The second aspect of rational thinking is *value judgment.* There is usually a great deal of room for argument here, even between persons purportedly espousing the same values, but especially between persons who fail to articulate their values clearly. The engineer can calculate the size of a bridge that could safely span the Grand Canyon, without much room for debate on the quantitative issues. But the question whether or not to put the bridge there in the first place is a value judgment that relates to why the bridge is there and to its impact upon other aspects of the setting. In the situation where a new power plant is proposed for a particular location, an environmentalist might claim that the ecology of the river will be upset. A biologist might be able to provide some numbers relating to the ecological impact (probably not very reliable because of poor or nonexistent data), and a heated debate over the quantitative issues could ensue. Various value judgments might be espoused: "That power plant will foul up our river." "The plant is essential to the economic survival of our community." Ultimately the decision should be made by considering the quantitative facts against a background of important values; the better understood the quantitative statements, the clearer the decision.

In this book we shall emphasize the quantitative aspects necessary for rational thinking in decisions about energy. The basic science of energy is simple, relates well to personal experience, and does not involve very complicated mathematics. If you do not already know about the simple scientific matters that we will discuss here, you may be amazed to find how much *you* will be able to do with some very simple tools for quantitative analysis. We expect, for example, that before long you will be able to make a fair estimate of the temperature rise of the river used by that new power plant that is proposed for your community, or to assess quantitatively the impact of a national emphasis on generation of electric power from solar energy. You may find that your ability to analyze these situations quantitatively is very satisfying, in which case you will experience some of the emotions of an engineer. You will probably find that your ability to communicate with your engineering friends will improve dramatically as you develop your ability to think and talk about matters of mutual interest. Very soon you will be able to read articles in such journals as *Scientific American*, *Mechanical Engineering*, *Power*, *Technology Review*, and the *Society of Automotive Engineers Journal* (are these in your library?). From these journals you will be able to extract quantitative information that you can understand and can use in your arguments for or against a particular issue. Please always bear in mind that our objective is to give you some appreciation for what is involved when an engineer does a quantitative analysis, *not* to make you a skilled engineer! If you already have this quantitative bent, and have studied math, science, or engineering, look for new ways to think about the scientific matters discussed here; you may find new understanding, and perhaps some simple ways to explain your life to your nontechnical friends.

"Science" is the systematic structure of ideas that are used to describe nature. The task of creating things that take advantage of scientific principles to do certain jobs is called "engineering." Thus, the engineer must understand and use science, and in addition have some inclination for the nuts-and-bolts aspect of making real things work properly, cheaply, safely, reliably, and responsibly. The technique of doing all this is called "technology." The technology of road building is relatively simple, and does not involve much science or mathematics. The technology of satellite communications is very sophisticated, and involves very newly developed scientific ideas (some discovered by engineers and some by scientists) and high-powered mathematics. Some engineers are attracted by the simplicity and effectiveness with which they can apply low technology, and others are attracted by the complexity of high technology. The technology of energy spans the spectrum from the relatively low (windmills on the farm) to high (magnetohydrodynamic power generation); so there is room for all kinds. Our objectives are to give you some background in energy science, some experience with energy engineering, and some feeling for energy technology.

To help augment your feeling for hardware and your understanding of energy technology, we suggest that you make arrangements to visit a variety of places nearby, preferably as you are studying something about that particular technology. Be sure to visit a large central power station. If you are on or near a university, visit their mechanical engineering laboratory, where you will probably find examples of engines, heat exchangers, turbines, compressors, boilers, etc. Or perhaps you can find a factory with a central heating and cooling plant, or a large air-conditioning system, where you can see things like cooling towers, pumps, etc.

During each visit get a quantitative feel for the size of the facility, say in terms of the amount of energy or power handled, try to learn the capital-equipment cost, and then calculate the cost per unit energy or power. Ask yourself what a larger unit would be like, what the impact of the facility is upon its community, and how such systems directly affect your life. You can do a lot on your own to gain more appreciation of energy technology; this book should be only the start.

Throughout the book we will raise questions that do not have unique answers, but for which a combination of quantitative analysis and value judgments provides the best basis for rational decisions. Try out your tools for quantitative analysis on these questions; try to identify the pitfalls in your analyses, and estimate the possible ranges of the quantitative results. And try to articulate the values that you feel are important to the questions and your own particular position on these values. Are your values shared by your society? By other cultures? How do the quantitative facts of life alter or influence your values and those of society? How could (or should) the values of your society be changed?

In this book we will concentrate on the technical aspects of energy. You must be the author of your values, but we will try to help you understand, clarify, and extend them. We hope that this experience will help you develop a style of critical thinking that will serve you in many other aspects of your life.

IT ALL ADDS UP

What is energy? A good portion of this book will be devoted to the job of developing an understanding of energy. As with all scientific concepts, one develops good understanding only over a long period of time, with frequent exposures from many angles. Think back; when did you first hear about energy? Perhaps in the early elementary grades, if not before. As a young child, you may have read breakfast-cereal boxes, where the contribution of the cereal to your minimum daily demands for energy was proudly announced. You may have been scolded for wasting energy by leaving on lights unnecessarily; through such experiences you became aware of the fact that energy costs money. You no doubt saw commercials describing how a particular high-energy gasoline makes cars perform better, and you perhaps visited a large power station near a dam on a summer's vacation. If you grew up on a farm, you soon learned the notion that machinery could do a lot more work than you yourself could do by hand. If you studied biology, chemistry, or physics in school, you were again exposed to the concept of energy. The totality of your exposure to various aspects and kinds of energy puts you in a very good position really to understand the concept of energy. Imagine how frustrated you would feel if you began a study of energy without this personal background! Our first job is to tidy up your understanding of the concept of energy; you have made this easy for us simply by living to your present age in a world in which energy plays such an important role.

There are three important aspects of the concept of energy; probably you already know all these, but you may not have heard them articulated. They are

1 All matter and all things have energy.
2 The energy of the whole is the sum of the energies of the parts.
3 Energy is conserved.

The first recognizes that energy is a property of matter; the molecules in a certain chunk of matter, the electromagnetic waves in a certain field of radiation, or the cells in a certain living organism have energy. The second aspect tells us that the amount of energy in a complex system is the sum of the energies in its various parts. This may seem trivial, but matter has other properties (for example, temperature) which do not behave in this manner. The third states that the energy of a system that does not interact in any way with anything else is constant, in other words, that the universe always contains the same amount of energy (though perhaps the *form* of the energy might change).* Why are these three statements true? You might think that they are experimental observations, but in fact they are really scientific *axioms*. We can't really tell anybody what energy is without invoking these three ideas; they are part of the basic concept of energy. They are as fundamental as energy itself.

Often we want to analyze systems that interact strongly with their surroundings, and to do this we use *energy bookkeeping*. As any good bookkeeper knows, we first must define the system that we are dealing with and the energy flows to and from it very carefully. (If you don't distinguish carefully between deposits and withdrawals, and make them for the proper bank account, you are likely to end up bouncing checks!) Engineers usually express their choice by drawing dotted lines around the part of the world that they wish to consider as "the system." Figure 1.1 shows a schematic of a large central power station in the process of burning a fossil fuel for the purpose of producing electric power. Not all the fuel energy can be converted into electrical energy, and some rejected energy leaves through the boiler stacks and some more through the

---

* We shall use the term "conservation" of energy only to reflect this idea, which should not be confused with the popular idea that "energy resources" must be conserved for the benefit of later generations.

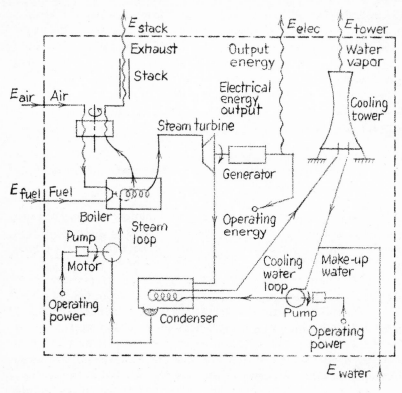

FIG. 1.1 SCHEMATIC OF A STEAM POWER PLANT.

cooling-tower exhaust. One possible choice for the "system" (sometimes called "the control volume") that would be analyzed in the study of the energetics of this power station is shown by dotted lines in Fig. 1.1. Figure 1.2 is a photograph of a real power station that uses ocean water for cooling rather than cooling towers. We have drawn a dotted line on this picture to show the equivalent system boundary. The three ideas discussed above tell us the following:

1 There is some energy in this system stored in the mass within; there is also a flow of energy into the system with the fuel, air, and water and out of the system with the stack and cooling-tower exhausts. And of course there is an energy outflow associated with the electric current.

2 Since energy is conserved, the energy that flows into the system during one hour must account for the energy that flows out during

that period plus any increase in the energy stored within the system. This is the statement of correct energy bookkeeping.

3 At any instant in time, the energy within the system is the sum of the energies of all the parts (when the plant is operating steadily, this energy would be the same at all instants in time, and this can lead to important simplifications in the energy analysis).

The second gives us the basis for writing down a simple equation describing the conservation of energy for this system, called the "energy balance,"

$$E_{\text{in}} = E_{\text{out}} + \Delta E_{\text{stored}} \tag{1.1}$$

Here E stands for energy; E_{in} represents the amount of energy that enters the system in the time period under consideration, E_{out} the amount that leaves the system in this period, and ΔE_{stored} the change in the amount of energy stored within the system. In the language of mathematics the symbol Δ is used to denote "final minus initial" (which of course is the "change"), and so

$$\Delta E_{\text{stored}} = E_{\substack{\text{final} \\ \text{stored}}} - E_{\substack{\text{initial} \\ \text{stored}}} \tag{1.2}$$

To do our bookkeeping properly, the three terms in Eq. (1.1) must of course be evaluated for the same time interval and in consistent measuring units. And we must include all significant inflows and outflows of energy in the evaluation of E_{in} and E_{out}. Thus, for Fig. 1.1, we might write

$$E_{\text{in}} = E_{\text{fuel}} + E_{\text{air}} + E_{\text{water}} \tag{1.3a}$$

$$E_{\text{out}} = E_{\text{elec}} + E_{\text{stack}} + E_{\text{tower}} \tag{1.3b}$$

thereby neglecting other energy flows that we decide are not significant for this system (such as the energy input from the sun).

What could we do with all this? If we can develop numbers for each of the terms symbolized above, we could first examine the system to see if indeed we had an "energy balance." An imbalance would not indicate that the conservation-of-energy notion is invalid but would instead indicate either that our numbers or arithmetic were in error or that we had neglected some significant terms. If the plant is operating steadily, the amount of energy stored within the control volume should be the same at all times. This "steady-state" condition would allow us to set

$$\Delta E_{\text{stored}} = 0 \tag{1.4}$$

in which case the total energy input must be precisely balanced by the total energy output. Since the objective of the power plant is to convert the energy in the fuel to electrical energy, we could quantify the ability

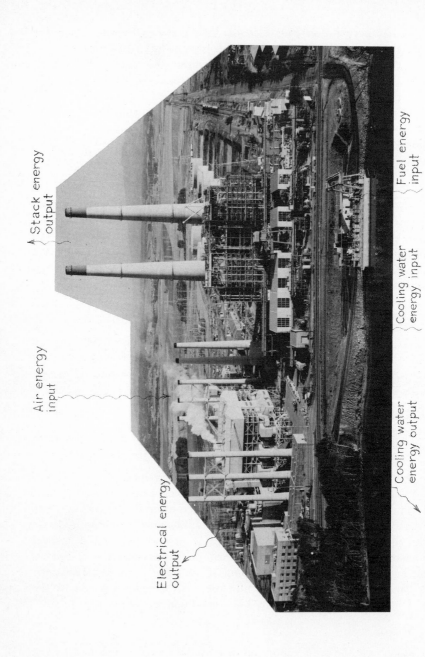

Stack energy output

Air energy input

Electrical energy output

Cooling water energy output

Cooling water energy input

Fuel energy input

FIG. 1.2 THE MOSS LANDING POWER STATION: 2133 Mw, FOSSIL-FUELED, OCEAN COOLANT. COURTESY OF THE PACIFIC GAS AND ELECTRIC COMPANY.

of the power plant to do this by defining an "overall plant efficiency," which normally is denoted by the greek symbol η ("eta"),

$$\eta = \frac{E_{elec}}{E_{fuel}} \qquad (1.5)$$

In a typical power plant perhaps 30% of the fuel energy would be converted to electrical energy ($\eta = 0.3$). We could then look at E_{stack} and E_{tower} to see where the rest of the energy is going. If we are clever, perhaps we could figure out some way to make use of this "waste energy." Or perhaps the values of the times would dictate that it is "best" simply to waste this energy.

The notion of an energy balance on a well-defined system, as illustrated above, is very important. It is the primary tool by which one derives quantitative information needed for the design, analysis, and decision making involved in energy systems. It is not very difficult; it is simply a matter of good bookkeeping (evaluating the various terms in the energy balance can be more tricky). You will learn how to do some simple energy problems yourself, and this will help you understand the harder ones.

GETTING USED TO NUMBERS

Let's begin to acquaint ourselves with numerical values of energy. There are several important measures of energy, or energy units, in common use today. These include

1   The *calorie*, which is roughly the amount of energy that must be added to 1 gram of water to increase its temperature by 1°C.*
2   The *Btu* (British thermal unit), which is roughly the amount of energy that must be added to 1 pound of water to increase its temperature 1°F.†
3   The *joule*, which is precisely the amount of energy that is used by a one-watt light bulb in one second.

One Btu corresponds to 252 calories; equivalences of these and other energy units are given in the Appendix, Table A.3.

You will want to develop some personal "feel" for the magnitudes of energy. One "calibration" point is that your daily intake of food energy

---

* °C stands for "degrees Celsius." You may know of it as "degrees centigrade," which is an older way of naming the same temperature interval.

† °F stands for "degrees Fahrenheit," as you probably know. An interval of 1°F corresponds to an interval of $\frac{5}{9}$°C.

is around 2 to 4 million (i.e., 2 to 4 × 10^6) calories,* depending upon your particular size, shape, and activity. In terms of Btu, this is of the order of 8,000 to 16,000 Btu.

Check the specification plate on your water heater at home; it probably says something around 10,000 Btu, which indicates the amount of energy that the heater is able to put into the water in one hour. Then, you can calculate how many pounds of water your heater can warm up 100°F in this period.

Another helpful reference is the "toaster" calibration. A typical toaster requires about 1,000 joules of electrical energy per second of operation. It takes perhaps 20 seconds to make tasty toast, or about 20,000 joules. A joule is about ¼ calorie (see Table A.3). So the amount of electrical energy required to cook two pieces of toast for your breakfast is about 5,000 calories, or about 0.2% of your body's energy requirement for the day.

If you are not familiar with the various energy units, do simple calculations like these as you study the subject, and relate the energy requirements or output of various devices to the "you" calibration and the "toaster" calibration; before long you will have a very good feel for the magnitude of energy numbers.

An important aspect of systems that use or produce energy is the *rate* at which this happens. A battery that could deliver 20 × 10^8 joules would be ideal for an electric urban automobile; but the battery would be useless unless this energy could be delivered in a few hours (nobody wants to take three days to go the the supermarket). The "power" of the device is the characteristic that describes the *rate* at which it supplies or uses energy; the power is the energy divided by the time required for the energy transfer. Denoting the power by P, the energy by E, and the time interval by t, the power is†

$$P = \frac{E}{t} \tag{1.6}$$

---

* 1,000 calories = 1 kcalorie (kilocalorie). Food values are usually stated in kcalories, often denoted simply as Calories (note the capital C that is often left out inadvertently by dietitians!).

† If one wants to get fussy with the mathematics, the time interval should be taken as very small, in which case the energy transfer is very small, but the ratio of the two (the power) has a finite value. Equation (1.6) applies for longer time intervals whenever the power as so defined is uniform over the time interval, and is also used to describe the *average* power over a long time interval.

For example, your calorific intake of 3,000 kcal in 1 day corresponds to an average of $P = 3,000$ kcal/24 hr $= 125$ kcal/hr, which corresponds to a steady diet of one cookie per hour. If you eat a 2,000-kcal dinner in 1 hr, your power input is 2,000 kcal/hr, 16 times the power input of a cookie nibbler.

The power used by a toaster is about 1,000 joules/sec. The unit combination *joules/sec* (joules per second; per means "divided by") is termed "watts" after a famous contributor to energy technology; 1,000 watts is termed a kilowatt ("kilo" = 1,000), abbreviated by Kw or kw. So, your toaster power is about 1 kw. Power units are often given in kw, which is why the "toaster calibration" is so useful. Another unit of power that is in common use is the *horsepower*, abbreviated hp or HP, after a famous horse. This is roughly the amount of power that a horse can deliver; it is also equivalent to about $\frac{3}{4}$ kw. In other words, a horse working hard could deliver enough power to cook your breakfast toast. (Are *you* more or less powerful than a horse?)

Since electrical power is usually measured in kw, electrical energy is usually given in kwhr ("kilowatt hours"; power times time equals energy). Check the electric meter outside your residence; read the amount of energy (in kwhr) used to date, and then read it again later and compute the average electrical power that was used over this time period. Most appliances, lights, etc., have power requirements in kw or watts stamped on them; check these to see if you can account for the energy that the electric company says you used. By the way, they probably charge you about 2.5 cents per kwhr (check your electric bill). You might compute the cost of making toast, or of leaving a light burning all night, to further increase your personal feel for energy and power.

The horsepower unit is used for motors, pumps, and other devices that produce or use energy. Get a feel for horsepower; remember that a kilowatt is about $\frac{4}{3}$ of a horsepower. Your kitchen refrigerator probably has about a 1-hp motor; your heart uses about 0.01 hp; your automobile engine may be rated at 300 hp but probably seldom actually produces more than 100 hp, even under drag-strip conditions. Large jet engines produce about 25,000 hp. The largest modern steam turbines in central power stations produce about a million horsepower.

Figure 1.3 gives some typical energy and power numbers for some interesting systems. Note that the range of interest is enormous, and hence a logarithmic scale has been used in each case. You will recall that multiplication is equivalent to adding logarithms; this means that a fixed percentage change is represented by a fixed interval on the logarithmic scale, irrespective of where on the scale the change is taken. Logarithmic

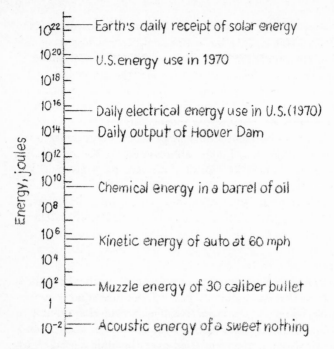

FIG. 1.3a THE ENERGY OF THINGS.

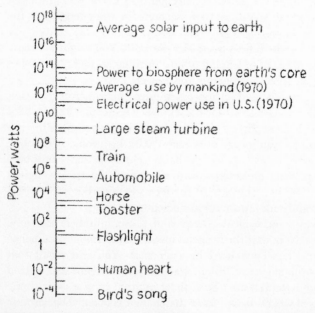

FIG. 1.3b THE POWER OF THINGS.

scales such as those used here are very helpful in studying phenomena involving rapid growth, such as the national consumption of energy; if they are new to you, buy some "log-log" paper and some "semilog" paper at your local bookstore, and play with these until you feel comfortable plotting or reading such graphs. It takes only courage, not mathematical wizardry!

At this point it is important that you work seriously with the ideas described above to get a reasonable feeling for energy and power magnitudes. You may not be quite sure just what energy is, but we'll clarify this in the next chapter. The remainder of this chapter provides an overview of the patterns of energy and power utilization in our contemporary society. A reasonable feeling for energy and power magnitudes will help make this material meaningful to you. You may also want to read it again after you have learned more about energy technology.

PATTERNS AND CHANGES

The United States, with 6% of the world's population, accounts for about 35% of the world's energy utilization. The United States *per capita* utilization of energy (all forms) was about 250 kwhr per *day* in 1970. That's 250 hours of toasting *per person per day*! This is about 5 times greater than the world average, and of the order of 20 times greater than that of an emerging industrial nation, such as India. Indeed, there is a strong correlation between "standard of living" (by whose values?) and the per capita energy consumption of various nations (see Fig. 1.4). High energy consumption appears to be a necessary (but not sufficient) ingredient for a high standard of living.

The 250 toasting hours mentioned above is somewhat misleading, for less than 10% of the energy used was electrical. The electrical energy consumption in the United States in 1970 was about 1.5×10^{12} kwhr (*one and one-half trillion hours of toasting!*), which amounts to about 21 kwhr per day per capita (slightly less than one continuous toaster for every man, woman, child!).

Electrical energy must be obtained from more primitive energy sources. Fossil fuels were the source of most of the electrical energy used by the world in 1970; less than 5% was produced in nuclear and hydroelectric power stations. The world's prime users of energy were therefore also prime consumers of oil, gas, and coal, and in most cases also major importers of these fuels. You may be surprised to learn, however, that Japan imported twice as much oil as did the United States in 1970, although the United States burned about three times as much oil as did Japan. But this picture may change dramatically as a result of new national energy policies.

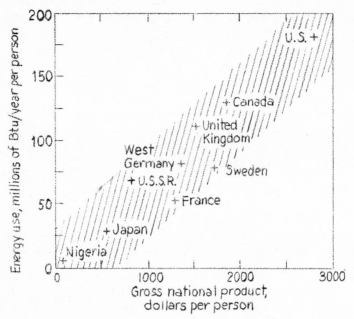

FIG. 1.4 ENERGY CONSUMPTION
AND GROSS NATIONAL PRODUCT.

Fossil fuels also provide a major source of energy for space heating and industrial-process heating. The energy equivalent of the fossil fuels burned in the United States during 1970 for all purposes was about 61×10^{15} Btu (which you can calculate is equivalent to 1.8×10^{13} kwhr). The breakdown was as follows:

Petroleum	24×10^{15} Btu
Coal	13×10^{15} Btu
Natural gas	24×10^{15} Btu
Total	61×10^{15} Btu

The energy equivalent of United States fossil-fuel imports (net) in 1970 was about 6×10^{15} Btu; in other words, the United States imported only about 10% of the energy that it utilized, taking practically all its needs from home sources. This is a direct result of basic policy decisions that were made by the Eisenhower administration. New policies might drastically change this distribution.

Figure 1.5 shows the utilization of energy in the United States in 1970. Note that about 18% of the energy was not really used at all but was just "wasted" in the process of producing and transporting electrical energy.

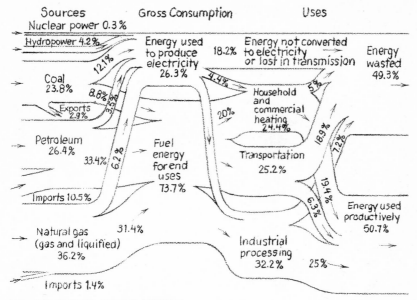

FIG. 1.5 UNITED STATES ENERGY
FLOWS IN 1970, EXPRESSED AS
PERCENTAGES OF THE GROSS
CONSUMPTION OF 64.6 × 10¹⁵ Btu.

As mentioned earlier, a typical power plant is able to convert only about
one-third of the input fuel energy into electrical energy, and the remain-
ing two-thirds makes up most of the "wasted" energy. Automobiles
accounted for the major portion of the energy used for transportation,
and space heating accounted for more than half the energy used in homes.
Industrial uses were quite varied; estimates suggest that the steel industry
was the largest single industrial user of energy, taking about 20% of the
energy used industrially.

The fractions of energy utilized by different sectors of the society have
changed somewhat over the years, but the striking change has been in the
amount of energy used each year by the society. Figure 1.6*a* shows the
total annual energy utilization in the United States since 1850, with
projections to the year 2000. Note that this is a linear scale. Figure 1.6*b*
shows the same information on a semilogarithmic scale. In regions where
a line is straight on a semilog plot the quantity in question is increasing
by a fixed fraction each year. We see that the utilization of energy in
the United States has been increasing by about 3.5% each year over the
past 25 years. In other words, it has taken only about 20 years for the
annual consumption of energy in the United States to double. We say

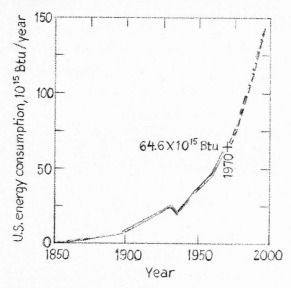

FIG. 1.6a ANNUAL ENERGY
CONSUMPTION IN THE
UNITED STATES (ALL FORMS).

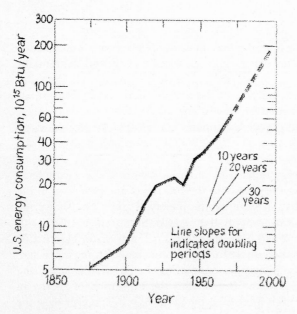

FIG. 1.6b ANNUAL ENERGY
CONSUMPTION IN THE
UNITED STATES (ALL FORMS).

that the energy consumption has a "doubling period" of 20 years.* This is indeed a frightening rate of growth, which is due in part to population growth and in part to technological changes ("technological advance" might be more appropriate, but this implies a particular set of values!). Figure 1.7 shows the per capita energy consumption in the United States over this period. Note that this was fairly constant during the period of the Industrial Revolution and has risen slowly since 1900, with a doubling period of about 50 years. (The doubling period for the United States population during this time was about 30 years.)

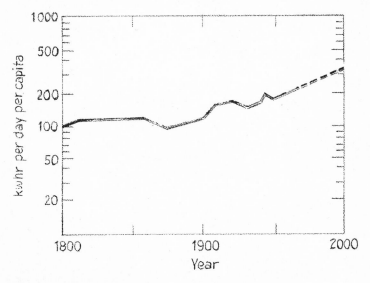

FIG. 1.7 CONSUMPTION OF ENERGY (ALL FORMS) BY THE UNITED STATES PER DAY PER PERSON.

One of the important energy sources is natural gas. Figure 1.8 shows the rate of consumption of natural gas in the United States. The straight line drawn for comparison has a doubling period of 10 years, which we see is about the doubling period for United States consumption of natural

---

* There is a frightening aspect about anything if its use doubles every so many years. Consider the series $1 + 2 + 4 + 8 + 16 + \cdots$. Note that each number is *more than the sum of all previous numbers*! Hence, in the period in which 16 of the quantities were used, *more were used than in all previous time*! Thus, every 20 years the United States consumes more energy than in the sum of all previous 20-year periods! Think about the implications of this "doubling law" on a population that doubles every 30 years.

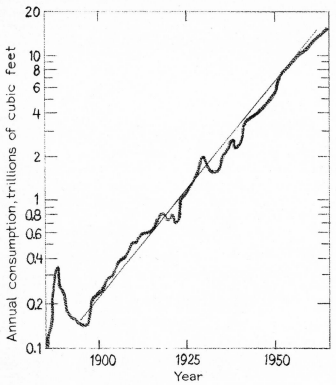

FIG. 1.8 CONSUMPTION OF
NATURAL GAS IN THE
UNITED STATES. THE LIGHT LINE
HAS A 10-YEAR DOUBLING PERIOD.

gas. Figure 1.9 shows the annual electrical energy output from central
power stations in the United States. These data display growth with a
doubling period of only 10 years! Most of the power was produced from
fossil fuels, and hence it is not surprising that the doubling periods for
natural-gas consumption and electrical-energy production are about the
same.

The history of the total energy utilization in the United States in the
period 1850–1970, with some extrapolations, is shown in Fig. 1.10.
Note that the time required for a new energy source to grow to the point
where it is a significant factor is of the order of 50 years. This is clearly
seen for coal, oil, and gas, and we are in the initial stages of the growth of
nuclear energy at the present time. One should learn from history;
what does this record suggest about the likelihood of converting the United
States energy system to a solar-energy base in a period of 10 or 20
years?

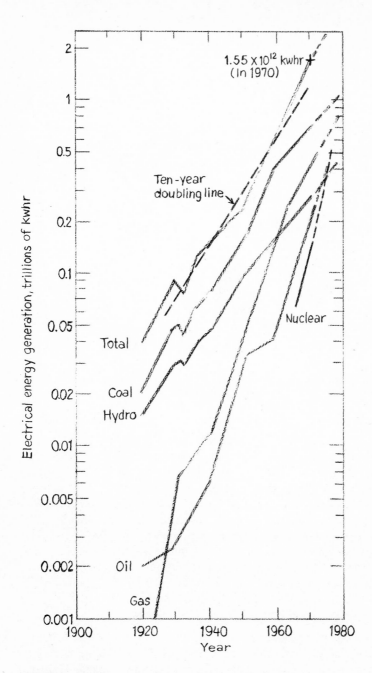

FIG. 1.9 HISTORY OF ELECTRICAL
ENERGY GENERATION IN THE
UNITED STATES.

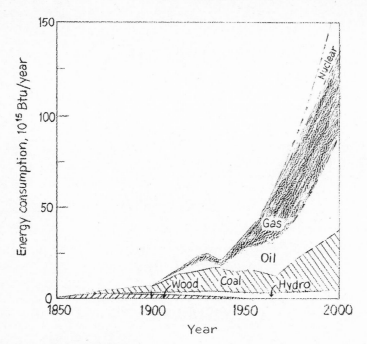

FIG. 1.10 CONTRIBUTION OF
VARIOUS ENERGY SOURCES TO
THE ANNUAL ENERGY
CONSUMPTION IN THE
UNITED STATES.

The history of the electric power industry in the United States is very
interesting. Whereas in some countries hydroelectric power once out-
ranked power generation from fossil fuels, and in some it still does, this
was never true in the United States. On Sept. 30, 1882, the country's
first hydroelectric station began operation on the Fox River in Wisconsin;
this plant had the capacity of 12,500 watts, only enough to light a few
hundred street lamps. However, 26 days earlier the first fuel-burning
power station was started up in Manhattan, with an installed capacity of
33,200 watts, to power a line of 400 light bulbs of 83 watts each. For a
long time hydroelectric plants contributed about 30% of the nation's
growing electrical-energy supply, but after World War II the importance
of hydroelectric power diminished. By 1970, the mix of electrical-energy
generation in the United States was about as follows:

Hydroelectric	16%
Fossil fuel	76%
Nuclear	8%
	100%

Figure 1.9 also shows the history of electrical-energy generation by hydroelectric power stations, various sorts of fossil-fuel-burning power stations, and nuclear power stations. In addition, industry generates some electrical energy for its own use—sugar mills burn the trash to make the steam that generates the energy that runs the mill that grinds the cane that makes the trash. Again note the small contribution of hydro-electric energy and the rapid growth in the generation of electrical energy from nuclear-energy sources. Indeed, the rapid expansion of the nuclear-power industry provided considerable stimulation for the ongoing heated debate over national energy policies.

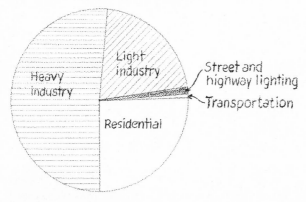

FIG. 1.11 APPROXIMATE DISTRIBUTION OF ELECTRICAL ENERGY IN THE UNITED STATES.

Where did all the electrical energy go? Figure 1.11 shows the uses of electrical energy in 1970. Residential uses account for about 35% of the electrical energy. Industry is the primary consumer of electrical energy (about 70%). We saw earlier that transportation was a major consumer of energy, but now we learn that only a small portion of electrical energy is used by transportation (most of the transportation system runs on fossil fuel and not electricity). This suggests that a serious effort at electri-fication of our transportation system (including automobiles) would have a drastic impact upon the distribution and consumption of electrical energy. Electrification of residential heating would also have a big effect and could easily quadruple the residential consumption of electrical energy. Continued growth of aluminum and titanium production, which uses gobs of electrical energy, could keep the industrial consumption large. Hence, it is not easy to tell precisely how the distribution of electrical energy will change over the next decades without first doing careful analy-ses of the important users of electricity. But the established trend of a

10-year doubling in the demand for electrical energy is likely to continue to the end of this century unless basic policy decisions force a change.

Earlier we cited the outputs of the first two United States electric power stations. Things have really changed; a large fossil-fuel-burning power station, such as the Pacific Gas and Electric Company plant at Moss Landing, Calif. (Fig. 1.2), is capable of 'producing around 2,000,000,000 watts of electrical power—two million kilowatts!—two million toasters! With so many zeros, the term megawatt (million-watt), abbreviated Mw (mw usually means milliwatt, or *thousandths* of a watt be careful!), is used. So plants today have installed capacities as large as 2,000 Mw.

Now that you know megawatts, you had better learn gigawatts, abbreviated Gw, which stands for 10^9 watts, or a thousand megawatts. Large plants have an installed capacity of a couple of gigawatts. Today there are of the order of 3,000 central power plants. Few are as large as 2,000 Mw; 350 Mw is a typical size for many of the more recent installations. If we divide the 1970 electrical-energy consumption of 1.5×10^{12} kwhr by 8,760 hr, which corresponds to one year, we can compute the average electrical power in the United States over the year as 1.7×10^8 kw, or 170 Gw. Now, the power-producing capability of the plants in the United States must be more than this, because not all plants operate at full power all the time (the demand for power varies). Indeed, the installed capacity in the United States in 1970 was about 300 Gw, slightly less than twice the average power for the year. If the 3,000-power-plant figure is about right, the "average" plant has an installed capacity of about 0.1 Gw, or about 100 Mw. This sounds about right; there are many older, smaller plants, and relatively few 1- and 2-gigawatt behemoths.

Until we find ways to alter the growing demand for electrical energy, or until nature limits the sources of energy, new power plants must be built each year. Westinghouse has estimated that between 1970 and 1990 the United States utility companies will have to add more than 1,000 Gw of new capacity; about half of this will represent the annual average, or "base load," which will be added to the 170-Gw figure that we calculated earlier. Much of this growth will be concentrated in nuclear power stations. In 1970, there were some 22 nuclear power plants in operation, 55 under construction, and 40 on order. Legislation that would delay the construction of these new plants has been seriously proposed and hotly debated; you can see from these numbers that the impact of such laws would be enormous, and no decision should be reached lightly or without careful consideration of all the pertinent issues.

Forecasting the nature of our energy-utilization patterns several decades ahead is currently fashionable. To do this accurately requires a very careful and complete analysis of such things as petroleum reserves, nuclear energy, fusion power potential, personal life styles, and the values of the society. At one time there was a scare that the fossil-fuel supply would be exhausted before the end of this century. Careful analysis suggests this is not the case but that our present rate of expenditures of fossil fuels, together with the projected growth in demands for energy, will indeed exhaust these sources in a matter of a few centuries. Without other energy sources, the technological age will be over before it has started, and will be almost invisible on a geological time scale. To illustrate the problem, let's suppose that we have now extracted 0.01% (one-ten-thousandth) of the earth's fossil-fuel reserves, and that the doubling period for consumption is 10 years.* A little calculation shows that in about 13.3 doubling periods, or 133 years, it will *all* be gone. Now, our estimate of 0.01% may be high (or low) by a large factor. Suppose instead we have so far used 0.000001% of the fossil fuels; with a 10-year doubling period, it will all be gone in only 266 years! Our great-great-grandchildren have a problem that *we* must help them solve *now*!

WHERE DOES IT COME FROM?

The original source of energy for us on earth is our sun. Let's apply the "energy-balance technique" to the earth; the system we shall analyze is shown in Fig. 1.12. The energy balance, taken over a period of 1 year, is

$$E_{in} = E_{out} + \Delta E_{stored} \tag{1.7}$$

E_{in}, the "income energy," is due almost entirely to the radiant energy received from the sun (moonlight and starlight are not significant, except as stimuli of interpersonal communication). The moon and sun provide a small but important part of E_{in} to drive the ocean tides. The earth,

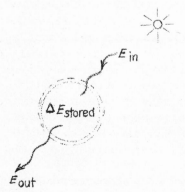

FIG. 1.12 ENERGY FLOWS FOR THE EARTH.

* The use of a fixed doubling period overemphasizes the difficulty.

being a fairly warm body, radiates energy to space. The atmosphere, the mantle, and the core of the earth are all changing temperature very slowly and undergoing chemical and nuclear changes (with our unsolicited help), and these changes appear in the energy balance as ΔE_{stored}.

Let's suppose for a moment that the sun suddenly went out, making $E_{\text{in}} = 0$. Then the energy balance would reduce to

$$\Delta E_{\text{stored}} = -E_{\text{out}} \tag{1.8}$$

Recall that means "final minus initial," so that ΔE_{stored} represents the change in the earth's energy over a year (the period of the energy balance), and Eq. (1.8) tells us that this must be a negative number. We could exist, for a time, without a sun, but the earth would quickly cool through the term E_{out}.

This discussion is somewhat misleading, because the rate at which energy can be transferred from the core of the earth to the surface, where we live, is very slow relative to human life processes. More insight can be obtained by analyzing just our surface layer, called the "biosphere." We have identified the important energy flows to and from the biosphere on Fig. 1.13. The energy balance on the biosphere, again over a 1-year period, is

$$E_{\text{sun}} + E_{\text{tides}} + E_{\text{geo}} + E_{\text{fuel}} = E_{\text{rad}} + \Delta E_{\text{biosphere}} \tag{1.9}$$

Here the terms represent, in order, the solar-radiation-energy input (E_{sun}), the tidal-energy input by the moon and sun (E_{tides}), the energy input from the earth's core by natural geological processes (E_{geo}), the energy input from the core by man's consumption of fuels (E_{fuel}), the energy radiated to space from the earth (E_{rad}), and the change in energy storage in the

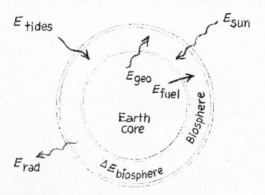

FIG. 1.13 ENERGY FLOWS FOR THE BIOSPHERE.

biosphere ($\Delta E_{biosphere}$). Some of these quantities have been measured quite precisely, and others can be estimated fairly closely. It is convenient to think in terms of the average powers for the year; we'll denote these by P_{sun}, P_{tides}, etc. Their values are as follows:

$$P_{sun} = 1.73 \times 10^{17} \text{ watts}$$

$$P_{tides} = 3 \times 10^{12} \text{ watts}$$

$$P_{geo} = 32 \times 10^{12} \text{ watts}$$

$$P_{fuel} = 6.3 \times 10^{12} \text{ watts (1970 figure)}$$

Now, the biosphere is very nearly in an energy balance, and $\Delta E_{biosphere}$ is very small (you can verify this by estimating the power that would be required to warm up all the earth's oceans at the rate of 1°C per hour, and compare this with the figures above). So, the output energy is very nearly in balance with the input, which is due almost entirely (at present) to the radiated solar energy. It is interesting to note that mankind already uses about twice as much energy as the moon and sun deliver to the ocean's tidal flow!

Let's see how far mankind can go without a serious impact upon the world environment. A large fraction of P_{sun} does nothing but keep the biosphere warm; about 23%, or 0.4×10^{17} watts, is used to evaporate water, drive the winds, etc., and we certainly don't want to make a substantive impact on that (one theory of the ice ages suggests that very small changes in E_{sun} and E_{rad} were responsible for these major changes in the earth's climate). Suppose it is possible to build up man's use of energy to an equivalent average power of 6×10^{15} watts. This is a factor of 1,000 over the 1970 value. Growth by a factor of 1,000 is obtained in only 10 doubling periods; for a 10-year doubling period, this is a span of only 100 years. In only about 150 years P_{fuel} would grow to the magnitude of P_{sun}, should the 10-year doubling period continue! Obviously we cannot continue, and must face up to the technical and political aspects of drastic changes in our energy utilization.

You should note that the energy balance on the biosphere [Eq. (1.9)] relates to energy flows to and from the biosphere. Any exchanges of energy that take place within the biosphere do not show up in this overall energy accounting. So, if we can find ways to capture some of P_{sun} early in the chain of energy transfer within the biosphere, use this energy for our purposes, and then return the energy to the biosphere, the overall balance will not be upset. If mankind really feels it needs to increase its energy consumption by a thousand-fold, which the 10-year doubling history says we will do in merely one century, then ways to use solar energy must be researched, developed, engineered, built, and placed into operation on a gigantic scale. It will cost a lot more than a dozen trips to the moon.

To summarize, all our energy ultimately came or comes from the sun. When we burn fossil and nuclear fuels, we use precious energy reserves that took millennia to create. If instead we can take advantage of income energy, we will have the opportunity for substantial but not unlimited increases in the energy available for our use. To know what to do, what positions to take in debates about energy, to know how to vote on issues relating to energy, we must develop a good understanding of these matters and of energy technology, and that is what the rest of this book is all about.

PROBLEMS

Getting used to numbers

1.1 A certain machine requires 24 kw of electrical power. How many Btu of energy does it use in an 8-hr working day?

1.2 A householder heater has an output rate of 100,000 Btu/hr. The homeowner wants to replace this with an electrical unit; determine the kw rating of an electrical unit that delivers energy at the same rate.

1.3 A residential watt-hour meter had the following readings:

April 22   19,273,427 watt-hr
May 22   19,383,792 watt-hr

Determine the amount of electrical energy, in kwhr and in Btu, used in this residence during this period, and the average electrical power.

1.4 A factory has the following power-time history:

8 A.M.–11 A.M.   790 kw
11 A.M.– 2 P.M.   360 kw
2 P.M.– 5 P.M.   1,370 kw
5 P.M.– 8 P.M.   460 kw
8 P.M.– 8 A.M.   120 kw

Calculate the daily energy expenditure, in kwhr and in Btu.

Forecasting

1.5 Starting with the 1970 United States electrical-energy consumption figure of 1.5×10^{12} kwhr, and assuming a 10-year doubling period for the subsequent 30 years, estimate the annual electrical-energy consumption for the period 1970–2000. Plot this on semilog paper, and read the predicted consumption for 1985.

1.6 Using the solution to Prob. 1.5, calculate the amount of electrical energy (kwhr) that will be consumed in the three 10-year periods of the 1970s, 1980s, and 1990s, and relate these to the "doubling law" discussed in Chap. 1.

1.7 Suppose the growth in the United States consumption of electrical energy can be reduced in a manner that doubles the doubling period every doubling period, with double the 1970 rate of 1.5×10^{12} kwhr being consumed in 1980, and then double this in 2000, etc. Forecast the annual demand for electrical energy on this basis, and compare the projection for 2050 with that calculated for Prob. 1.5.

1.8 Using the forecast of Prob. 1.5, how many 1,000-Mw power plants will have to be constructed between 1970 and 2000 to meet the projected demand at the turn of the century?

1.9 Using the forecast of Prob. 1.7, how many 1,000-Mw power plants will have to be constructed between 1970 and 2000 to meet the projected demand at the turn of the century?

1.10 Suppose 0.1% of the income energy from the sun can be harnessed for use by man, and that in the brave new world of A.D. 3000 each nation's share is allotted in proportion to its land mass. What would be the United States annual energy share (in Btu and kwhr) under this arrangement, assuming that the United States is the same size in the brave new world? Compare this with the current expenditure and with the year 3000 forecast based on the assumed 20-year doubling period.

Using the library

1.11 A good university library will contain a great deal of information relating to worldwide energy consumption. Locate the major sources of such information in your library, and prepare a short annotated bibliography identifying these sources and indicating the sort of information available in each.

1.12 Locate the "Minerals Yearbook," vols. I–II, published by the Bureau of Mines. Look up the annual demands for petroleum energy in Btu (e.g., page 28 in the 1967 volume), and plot on semilog paper for the period 1940 to the present. Determine the doubling period at present, and compare this with information cited in Chap. 1.

1.13 Using the source for Prob. 1.12, plot the annual use of petroleum energy for transportation since 1940, relate this to information presented in Chap. 1, and discuss implications.

1.14 Do Prob. 1.12 for coal.

1.15 Using the data source of Prob. 1.12, make a semilog plot of the annual consumption of a mineral of your choice, determine the doubling period, and discuss the relationship of this to the material in Chap. 1.

Keeping the energy books

1.16 Consider a household refrigerator. Draw a "system" diagram, showing the important energy flows that occur when the refrigerator is in operation. Write the energy balance in simple terms. What happens to the electrical energy put into the system?

1.17 Consider an idling automobile engine in a car standing still. Draw a "system" diagram and show the energy flows. Write the energy balance in simple terms. If the engine is operating at "steady-state" conditions, what happens to the energy that comes in with the fuel?

1.18 Draw the system diagram for an electric coffee pot, and show the energy flows. Write the energy-balance equation in simplest terms. Does your energy balance include enough terms to explain what happens when the pot is turned off?

1.19 Write an energy balance on yourself over a period of one day, using an appropriate "system." What happens to the energy that you take on with food?

1.20 An irrigation pump circulates water from the bottom of a hill to a small reservoir at the top. Define an appropriate system, write the energy balance, and relate the electrical energy input at the pump motor to the change in the energy of the water moved by the pump. Why is the motor needed?

2

IT ALL STARTS
WITH MECHANICS

In which we learn some of
the science of energy and
something about how to use it

THE CONCEPTION AND QUANTIFICATION OF FORCE

It is time now to start being more precise in our thinking about the science
of energy. If you are an engineer or scientist, you may already have a
good grasp of the material that we will cover in the chapter. Please read
it anyway—you may find some simple new ways to explain things in
your life to your nontechnical friends!

In order to understand energy, one must first understand the concept
of *force*. Think by yourself for a moment, and try to give a definition for
"force" . . . (these dots represent your thoughts). If you were able to
write down what to you seems an understandable definition, you have
somehow deluded yourself. Force, like any other concept, is something

that cannot be precisely defined but must be understood from the start.

The statement that "a force is a push or a pull" helps one get a feeling for force, but it is not an airtight definition; what, after all, is a push but a particular type of force? Pulls and pushes tend to move things (more precisely to *change* the motion of things), and so forces must be useful in explaining motion. Pushes are the opposite of pulls, and so forces must have some sort of directional characteristics. A force must have a point of application and certainly a magnitude. But these are only some of the features of force.

Figure 2.1 will help us to understand another aspect of the force concept. The man on the diving board in Fig. 2.1a pushes the board down a certain distance. The large rock on the board produces the same deflection. It makes no difference to the board whether the man or the rock is there; the deflection d is the same. We can remove the man and replace him by the rock, or we can replace him by a "force" F, which measures the strength of his interaction with the board. In Fig. 2.1c we show this replacement; the dotted lines show the "system" on which we are focusing, which excludes whatever it is that produces the force. So, another character of force is that it is something that we use to replace the interaction between a system and its environment when we mentally remove the environment to focus on the system.

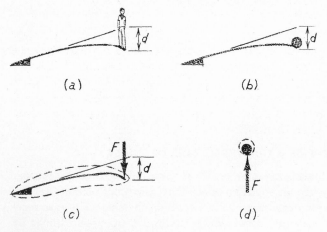

(a)

(b)

(c)

(d)

FIG. 2.1 FORCES ARE USED TO REPRESENT THE PUSHING OF SOMETHING ON SOMETHING ELSE. (a) THE MAN PRODUCES A CERTAIN DEFLECTION. (b) THE ROCK PRODUCES THE SAME DEFLECTION. (c) EITHER CAN BE REPLACED BY A FORCE. (d) AN EQUAL BUT OPPOSITE FORCE ACTS UPWARD ON THE ROCK.

The board also exerts a force on the rock. If we focus on the rock (Fig. 2.1*d*), we must show this force pushing *upward* on the rock. This force will be equal in magnitude but opposite in direction to the force that the rock exerts on the board. "Action equals reaction," perceived Isaac Newton. Is this statement an experimental fact, or is it part of the concept of force? If you think this question through, you will find that it is impossible to make an experimental test of the statement until you know how to quantify forces, and you can't do that without invoking the statement itself! If you push on a wall, you exert a force on the wall, and the wall exerts the same force on you. The notion that action equals reaction is part of the *concept* of force (see Fig. 2.2).

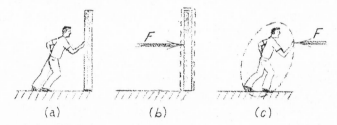

FIG. 2.2 HAVE YOU EVER BEEN PUSHED BY A WALL? (*a*) YOU PUSH ON A WALL. (*b*) YOU EXERT A FORCE *F* ON THE WALL. (*c*) THE WALL EXERTS AN EQUAL BUT OPPOSITE FORCE ON YOU.

Let's consider an object on which two forces act (Fig. 2.3*a*). The force F_A will try to produce acceleration (motion with increasing velocity) to the right, and F_B will try to produce acceleration to the left. If the object is not moving, it will tend to start moving in the direction that it is pushed by the larger force. The body will remain motionless in spite of the two forces only when the two forces are equal, i.e., if $F_A = F_B$. We say that the forces are "balanced." The notion that forces must be "balanced" on an object that is not accelerating is another part of the *concept* of force. Moreover, the forces can be "added," with due regard for direction, to determine the "net" force on the object. As we shall see momentarily,

FIG. 2.3 UNBALANCED FORCES PRODUCE ACCELERATIONS. (*a*) THE OBJECT WILL ACCELERATE TO THE RIGHT IF $F_A > F_B$. (*b*) THE OBJECT WILL NOT ACCELERATE IF $F_A = F_B$.

it is these aspects of the concept that enable us to quantify forces in a rational way.

"Weight" is the downward force that the earth's gravity exerts upon matter. An upward force of equal magnitude is exerted by the matter on the earth. If we define you as a system (Fig. 2.4), we replace the earth by two forces. First, we show the gravitation force W acting down on you, and then we show the earth pushing up on your feet with force F. If you are off your feet, the force W will try to pull you down. When you jump, you make F larger than W, and hence accelerate yourself upward until your feet leave the ground.

FIG. 2.4 FORCES THAT THE EARTH EXERTS UPON YOU. TO JUMP, $F > W$.

The concept of force becomes really useful when we can put numbers to forces. One might be inclined to use a force-measuring scheme such as that shown on Fig. 2.5. An unknown force is applied to the spring scale, and the amount of the force is read from the scale. The value of 1 on the scale would be determined by some standard force, such as the weight of a brass brick, and the other numbers would be distributed *linearly* along the scale. The trouble with this scheme is that it depends on the particular choice of spring; a brass spring and a steel spring could be made to agree at 1 but would give different values for larger or smaller forces. Besides, it seems unreasonable that the measurement of something as fundamental as force should depend on the whims of a spring-chooser. A better scheme is shown in Fig. 2.6. Again a reference spring is used, but this time it is used in only one deflected position. First, we place a mark at the point where the force equal to the weight of standard brass

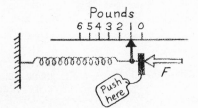

FIG. 2.5 AN UNSCIENTIFIC WAY TO QUANTIFY FORCE.

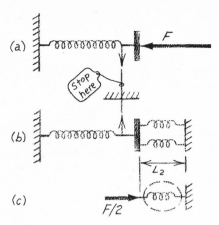

FIG. 2.6 A SCIENTIFIC WAY TO
QUANTIFY FORCE.

block deflects the spring by a certain amount; again this is our "unit force." Next, we find two perfectly matched springs (we can test this by observing that they have identical deflections under the same force). Then, we press these two springs against our reference spring, until the unit position is obtained (Fig. 2.6b). By symmetry, the force exerted by each of the two springs on the right must then be one-half of our unit force (which aspects of the force concept did we use here?); we can mark the deflection of one of these springs and use it to measure a half-unit force. We can continue this process and get springs to measure forces of one-third, one-twenty-ninth, etc., of our unit force; in principle we can collect enough springs to measure any magnitude of force to any desired degree of accuracy. In so doing we establish a *unique scale for force* that is fully consistent with the concept of force. This scale requires only *one* standard point (the weight of the standard brass brick) to set the values for *all* points on the force scale. While this is not the scheme used in practice, the schemes that are used are all based on the same basic ideas. The important thing to realize is that there are truly fundamental ways to quantify all scientific concepts, such as force.

Now you should begin to get the feeling that force has some very fundamental basis. Perhaps you also have the sense that science tries to think very fundamentally and very clearly about such concepts, and that your ability to use these concepts will be greatly enhanced when you do, too. If so, you have our message.

There are several units of force in common use today. You may be most familiar with the pound, which we shall denote in this book by lbf (pounds-force) to distinguish it from a mass unit lbm (pounds-mass) that we will encounter shortly. A pint of water weighs about one lbf ("a pint's a pound the world around"); since you are mostly water, if you know your weight in lbf, you can now estimate your volume in pints! Other force units and their equivalences are given in the Appendix.

A great deal of engineering involves consideration of forces. For example, suppose one wonders whether a certain pile of bricks can be lifted with a given cable. One first finds out the safe lifting load in lbf for the cable, then the weight of the bricks, and then makes the decision based on these two numbers. Engineers make many choices like this (most more complicated) in the design of houses, bridges, automobiles, and space vehicles.

The strength of materials depends upon the nature of the force. Figure 2.7 defines "tension," "compression," and "shear," the three main classifications of forces. Tension forces tend to elongate material, compression forces tend to squash it, and shearing forces tend to skew it. Note that tension and compression forces act *perpendicular* to the material surface, and hence these are called "normal" forces (meaning perpendicular and not non-queer). As you live the next few days, look around you for examples of normal, shear, torsion, compression, and tension forces to see how they arise and how they are balanced or used in a well-engineered system.

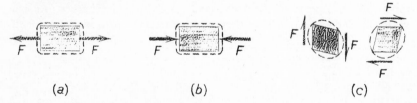

(a)   (b)   (c)

FIG. 2.7 TYPES OF FORCES.
(a) TENSION, (b) COMPRESSION,
(c) SHEAR.

To determine if a cable will break, an engineer will really be more interested in the "stresses" in the cable. Stress is *force per unit area*, and it is the value of the stress that really determines the deformation and failure of materials. For example, if the cable has a cross-sectional area of 0.5 square inch (scientific types write this as 0.5 in²) and carries a tension force of 15,000 lbf, the stress in the cable fibers is

$$s = \frac{F}{A} = \frac{15,000 \text{ lbf}}{0.5 \text{ in}^2} = 30,000 \text{ lbf/in}^2$$

The engineer would then look up the properties of the cable material to see if it could handle a tensile stress of 30,000 lbf/in². Most handbooks give the failure stress in "psi"; so you should know that "psi" stands for "lbf/in²."

In dealing with fluids (gases and liquids), the term "pressure" is used to describe the normal stress. Pressure is a compressive stress that acts perpendicular ("normal") to any surface in the fluid. The pressure is independent of the orientation of the surface. If we define a "system" to be a very small cube of the fluid (Fig. 2.8), then the compressive force is the same on each face of the cube, regardless of how it is tilted. The pressure is this compressive force divided by the area of a cube face.

FIG. 2.8 IN A FLUID THE NORMAL FORCE PER UNIT AREA IS INDEPENDENT OF DIRECTION.

The pressure of the atmosphere is about 15 lbf/in² (14.7 psi is closer). Hence, the average individual receives a rather substantial push in the belly from the air in front of him; fortunately this force is balanced by the pressure force on his back, and hence he is not accelerated by an unbalanced force (except perhaps on a very windy day when the pressure forces do not quite balance).

A piston engine (see Fig. 2.9) works because the pressure in the cylinder is greater than the pressure in the crankcase (which is usually about atmospheric pressure). This means that the downward force exerted by the gas in the cylinder on the piston is greater than the upward force exerted on the bottom of the piston by the crankcase gas, and this unbalance causes the piston to move downward. This in turn causes the piston to push on the connecting rod, to turn the crankshaft, to turn the driveshaft, to do useful work. The gas pressure in a typical automotive-engine cylinder might be 200 lbf/in² and typical piston area 20 in².

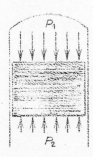

FIG. 2.9 THE PRESSURE IN THE CYLINDER IS GREATER THAN THE CRANKCASE PRESSURE; THIS PUSHES THE PISTON DOWN.

The force exerted by the gas on the piston would then be

$$F = PA = 200 \text{ lbf/in}^2 \times 20 \text{ in}^2 = 4{,}000 \text{ lbf}$$

The mechanical parts must be carefully designed to carry such large forces.

A bit of insight to the peculiarities of pressure gaugery may help your understanding of energy systems that you see. The notation "psia" is used to denote lbf/in^2 "absolute pressure," which is the true normal force per unit area. Most pressure gauges do not read pressure directly; rather, they read the amount by which the pressure is larger than the atmospheric pressure around the gauge. The notation "psig" ("psi gauge") is used to denote pressure levels *above atmospheric* pressure. Thus, a reading of 50 psig would actually correspond to an absolute pressure of $50 + 15 = 65 \text{ lbf/in}^2$. Most gauges are not carefully marked; many say "pressure in pounds" when they mean pounds per square inch above atmosphere. Others say psi when they mean psig. Barometers, of course, read the absolute pressure of the atmosphere. This is usually evaluated in "inches of mercury" or "millimeters of mercury," which refers to the height of a column of mercury whose weight would produce the atmospheric pressure at the base of the column. Typical atmospheric pressures are around 30 inches of mercury. Pressure gauges for vacuums (pressures less than atmospheric pressure) usually read "inches of mercury vac."; these indicate the amount that the test pressure is *less* than atmospheric pressure, expressed in terms of the pressure under an equivalent column of mercury.

WORK

Chances are you have at one time pushed a stalled automobile; it might be easy to push the car a few feet, but you would be exhausted if you pushed the car a mile. The force that you need to exert on the car is the same regardless of how far you push (on level ground it is determined by the rolling friction), but the longer the distance the more tired you get; the farther you push, the more energy you expend.

Let's consider the energetics of the car-pushing process. In Fig. 2.10a we show your pushing force F acting on the car (there may be other forces on the car as well, but we shall focus just upon F). Suppose the car moves a distance x in the direction of the force. The product of F and x is a measure of the effort expended by F in contributing to the motion of the car; this product is defined as the "work done" *by* the force *on* the car; denoting the work by W,

$$W = Fx \qquad\qquad (2.1)$$

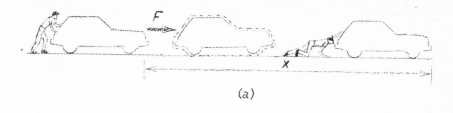

FIG. 2.10 (*a*) PUSHING A CAR =
HARD WORK; WORK = *Fx*.
(*b*) ENERGY DIAGRAM FOR
THE PROCESS.

The harder the push for a given distance, the more the work; the greater the distance for a given force, the greater the work.

Since work is a measure of effort, it must be a measure of energy. Indeed, the notion of work plays a crucial role in the structure of energy science; *work provides the fundamental way that the concept of energy is made quantitative*. When we say that the force *F* "did some work" on an object, we really mean that "some energy was transferred" from the thing producing the force to the object that moved (you expended energy to push the car). The amount of energy transfer associated with this process is *defined* by Eq. (2.1). Work, then, is an *amount of energy transfer*.

For example, if you push with a force of 70 lbf and move a car a distance of 200 ft, you will "do work on the car," or equivalently "transfer energy to the car as work," in the amount

$$W = Fx = 70 \text{ lbf} \times 200 \text{ ft} = 14,000 \text{ lbf-ft} \tag{2.2}$$

Since work is energy transfer, energy and work have the same dimensions (force times distance) and in the same units (lbf-ft, usually written as ft-lbf).

The energy balance picture for the car-pushing operation is shown in Fig. 2.10*b*. The energy input term is the work *W*; assuming there are no energy outflows, this must be balanced by the increase in energy storage. The energy balance is then

$$W = \Delta E_{\text{car}} \tag{2.3}$$

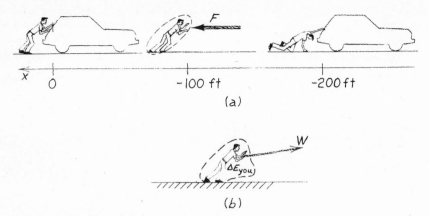

FIG. 2.11 (*a*) YOU MOVE IN THE OPPOSITE DIRECTION TO THE FORCE EXERTED ON YOU BY THE CAR. (*b*) SO, THE CAR DOES "NEGATIVE WORK" ON YOU, I.E., YOU DO POSITIVE WORK ON THE CAR.

This states that the energy that you put into the car as work must show up as increased energy within the car. Part of this increase would be in the energy of motion of the car; the rest would be in the increased energy of the tires, which will get hotter as a result of friction.

Let's turn the car problem around, and focus on you as a system (Fig. 2.11). Now we must be careful; in the definition of work, x is the distance that the matter on which the force acts moves in the direction of the force. In Fig. 2.11 we show a coordinate axis pointing in the direction of the force that the car exerts upon you; as the car moves 200 ft to the right, this coordinate changes from 0 to *minus* 200 ft. So, the distance that you move in the direction of the force acting upon you is -200 ft, and the energy transfer as work *to* you *from* the car is

$$70 \text{ lbf} \times (-200 \text{ ft}) = -14{,}000 \text{ ft-lbf} \tag{2.4}$$

Note that the work done *on* you is negative; this means the work done *by* you is positive, as common sense indicates. We denoted the work done *by* you as W. Then, the energy balance on you (Fig. 2.11*c*) is

$$\underset{\substack{\text{energy} \\ \text{input}}}{0} = \underset{\substack{\text{energy} \\ \text{output}}}{W} + \underset{\substack{\text{increase in} \\ \text{energy storage}}}{\Delta E_{\text{you}}} \tag{2.5}$$

Recall that $\Delta E_{\text{you}} = E_{\text{you final}} - E_{\text{you initial}}$, so that Eq. (2.5) says

$$E_{\text{you final}} = E_{\text{you initial}} - W \tag{2.6}$$

Thus, the energy stored within you *decreases* by an amount of W, and the energy stored in the car *increases* by an amount W. It all makes sense, doesn't it?

What happens if we take a third system, consisting of both you and the car (Fig. 2.12)? Now the force between you and the car is *internal* to the system; it will not appear in the system energy balance, just as the forces in bolts in the car or the forces in your bones and muscles did not appear before. *The only forces that can result in energy transfer across the boundaries of the system are those forces which act across the boundaries of the system.* This explains why engineers are so fussy about using dotted lines to carefully show their system boundaries; it's a good practice that makes for clear, accurate thinking.

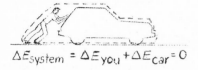

$$\Delta E_{system} = \Delta E_{you} + \Delta E_{car} = 0$$

FIG. 2.12   YOUR PUSH IS INTERNAL TO THIS SYSTEM, AND DOES NOT CHANGE THE SYSTEM ENERGY.

Have you been worried about other forces on the car or on you? There is certainly a large weight force acting downward on the car and a large upward force on the tires. Neither of these forces does any work on the car, because the car does not move in the direction in which these forces point. The same goes for the vertical forces on you. But what about the horizontal traction forces on your shoes and on the tires? Unless you kick up dirt, you really don't move the earth and hence do not do work on the earth with your feet. So no energy-transfer term need be shown at your feet. Similarly, unless the car is skidding, the tire surface on the road is at rest with respect to the earth, and hence the matter on which the roadway traction force acts does not move in the direction of this force. So, there is no energy flow term required there either. *Work is done by forces only when the material on which those forces acts moves in the direction of the force.* Otherwise, there is no work done and no energy transfer.

We have emphasized that work is determined by the motion of matter within the system. Motion is a relative thing, and so therefore is work. Careful energy analysis requires clear choice and definition of the "reference frame." For example, in a reference frame attached to your moving car the car has *no* motion. Hence, a rider in your car could claim that you may have pushed like crazy, but relative to him you did no work on the car! However, he should commend you for your efforts at pushing the earth backward, and would compute the amount of energy that you pushed into the earth as 14,000 ft-lbf. In his reference frame he would see an energy inflow to the car from the earth at the tires, and this is what he would use to explain the increase in energy stored within the car. Good energy bookkeepers get the same quantitative results regardless of the reference frame that they choose for their analysis. Considerate bookkeepers take the trouble to tell the auditors what they are doing; so be sure to define the reference frame if there is any possible question.

How would you go about evaluating work when the force varies as the object moves? Figure 2.13 shows the way. One divides the motion up into a number of small steps, computes the amount of energy transfer as work for each step, treating the force as constant over each small step, and then adds these up to get the total amount of energy transfer as work. You can see that the total work done by the varying force F as the object moves from point a to point b is approximately the area under the curve, and if the little steps in x are taken small enough, the sum of the work steps will become exactly the area under the curve. If you have studied calculus, you will recognize this process as "integration."

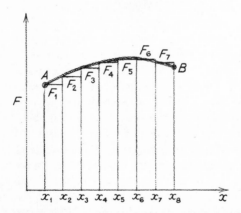

FIG. 2.13   THE TOTAL WORK DONE IN GOING FROM A TO B IS THE AREA UNDER THIS CURVE: $W = F_1(x_2 - x_1) + F_2(x_3 - x_2) + \cdots$.

Recall that power is the time rate of transfer of energy. Henceforth we shall use a dot over a symbol representing energy to denote a *rate* of energy flow; so $\dot{W}$ will represent the rate of energy transfer as work, i.e., the *power*. If it takes 2 min to push the car 200 ft with the 70 lbf push, then the power that is being delivered to the car during this period is

$$\dot{W} = \frac{W}{t} = \frac{14,000 \text{ ft-lbf}}{2 \text{ min}} = 7,000 \text{ ft-lbf/min}$$

One horsepower is *defined* as 33,000 ft-lbf/min; so you delivered about $7,000/33,000 = 0.2$ hp. This is a good figure to remember; at your best you are about a quarter horse! You can look in the Appendix for a conversion factor, and convert your 0.2-hp capability to kw.

To summarize, work proves the fundamental quantitative measure of energy. The amount of energy transfer to a system by a force that acts on the system is the product of the force and the distance that the matter on which the force acts moves in the direction of the force. Only forces that act across the boundaries of the system are involved. Good practice involves careful and clear definition of the system and the motion reference frame when one is analyzing energy flows as work.

Let's look at a simple engine to bring the ideas discussed above into sharper focus. The engine, shown in Fig. 2.14, uses high-pressure air as the "working fluid." When the piston is at its uppermost position ("top dead center," or TDC) the intake valve opens and high-pressure air rushes into the cylinder. This high pressure pushes the piston downward, and the intake valve remains open to let more air in during this downward thrust. When "bottom dead center" (BDC) is reached, the intake valve is closed and the exhaust valve opens, allowing the high-pressure gas to escape to the atmosphere, reducing the pressure in the cylinder. The inertia of the rotation then carries the piston upward, expelling more air through the exhaust valve. At TDC the exhaust closes and the intake opens, and we begin again. The pressures that might exist in the engine are shown in Fig. 2.14*b*; Fig. 2.14*c* shows a simplified approximation to the cylinder pressures that we will use to calculate the energy delivered to the crankshaft per cycle of operation. We'll assume that the piston diameter is 4 in and that the stroke is 6 in. Here followeth the engineering calculations:

Piston area:

$$A = \frac{\pi D^2}{4} = \frac{\pi}{4} \times (4 \text{ in})^2 = 12.5 \text{ in}^2$$

Force on piston during downward stroke:

$$F_1 = P_1 A = 200 \text{ lbf/in}^2 \times 12.5 \text{ in}^2 = 2,500 \text{ lbf}$$

Work done by high-pressure gas on piston during the downward stroke:

$$W_1 = F_1 x = 2,500 \text{ lbf} \times 0.5 \text{ ft} = 1,250 \text{ ft-lbf}$$

Force on piston during upward stroke:

$$F_2 = P_2 A = 20 \text{ lbf/in}^2 \times 12.5 \text{ in}^2 = 250 \text{ lbf}$$

Work done by piston on low-pressure gas during the upward stroke:

$$W_2 = F_2 x = 250 \text{ lbf} \times 0.5 \text{ ft} = 125 \text{ ft-lbf}$$

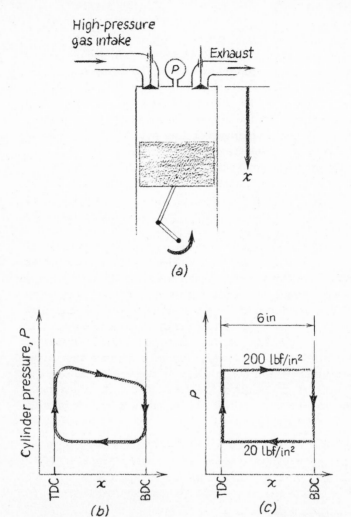

FIG. 2.14 ANALYSIS OF A SIMPLE ENGINE. (a) ENGINE DESIGN, (b) CYLINDER PRESSURE CYCLE, (c) SIMPLIFIED CYLINDER PRESSURE CYCLE.

Net energy transfer as work to the piston in one cycle:

$$W_1 - W_2 = 1,250 \text{ ft-lbf} - 125 \text{ ft-lbf} = 1,125 \text{ ft-lbf}$$

Assuming that the crankcase pressure is constant, there is no net energy transfer between the crankcase gas and the piston over one cycle. Suppose now that the engine runs at 2,000 rpm ("revolutions per minute") and that there are six such cylinders. The power delivered to the pistons, which except for friction will be the engine output power, is then

$$\dot{W} = \frac{1{,}125 \text{ ft-lbf}}{\text{cyl-rev}} \times 2{,}000 \text{ rev/min} \times 6 \text{ cyl} = 13{,}500{,}000 \text{ ft-lbf/min}$$

Converting to horsepower (see the Appendix for the conversion number),

$$\dot{W} = \frac{13{,}500{,}000 \text{ ft-lbf/min}}{33{,}000 \text{ ft-lbf/min-hp}} = 409 \text{ hp}$$

Engineers delight in making simple, useful calculations like these; if you understand the concepts of force, work, and energy, you can share in these pleasures. Calculations like these, but much more complex, are an essential step in the design of any power system.

NEWTON'S LAW OF MOTION

We have noted that forces tend to accelerate objects on which they act. The "mass" of the object is a measure of its ability to resist this acceleration. Under the action of a given force, a massive body will undergo a slower acceleration than a lighter body. The "mass" M of an object is defined by the relationship between forces on the body and accelerations that they produce; this relationship is known as "Newton's second law of motion." In physics books it is written as

$$F = Ma \tag{2.7}$$

Here F is the force on the object, M is the mass of the object, and a is the acceleration (time rate of change of velocity) that results when F is the only force acting upon the object. If we want to view Eq. (2.7) as a defining equation for mass, which is sometimes the viewpoint adopted, then*

$$M = \frac{F}{a} \tag{2.8}$$

In this view, to find the mass of an object one would apply a known force, measure the acceleration, and then simply divide one by the other. The dimensions of mass are then those of force divided by acceleration, and the units then depend upon the units of force, length, and time. For example, if F is measured in lbf, and a in ft/sec^2, then M has the peculiar units of lbf-sec^2/ft. This combination has been termed the "slug," and so by

---

* The mass of a body is independent of its speed of motion, except at velocities approaching the speed of light (186,000 miles/sec). At such incredible speeds one needs to revise his concepts of length and time, and this was the unique contribution of Albert Einstein.

definition 1 slug $= 1$ lbf-sec^2/ft. For example, if a force of 200 lbf produces an acceleration of 4 ft/sec^2 on an object, the mass of the object is

$$M = \frac{F}{a} = \frac{200 \text{ lbf}}{4 \text{ ft/sec}^2} = 50 \frac{\text{lbf-sec}^2}{\text{ft}} = 50 \text{ slugs}$$

Another important point of view, which eventually will be universally used, is the so-called "metric system." Here mass, not force, is viewed as being more fundamental, and a standard for mass (but not for force) is established. This standard is the kilogram, which is the mass of a particular chunk of platinum kept in a safe place in France. The units of length and time are the meter (which is defined in terms of the radiation wavelength of a particular element, but you can think of it as being the length of a platinum rod in that same safe spot in France) and the second (see if you can find out what the standard for time is yourself!). Then, in the mks ("meter-kilogram-second") metric system the dimensions of force are those of mass times acceleration, i.e., mass times length divided by time squared, and the units are kg-m/sec^2. This combination of units is called a "newton,"

$$1 \text{ newton} = 1 \text{ kg-m/sec}^2$$

and so the newton is the force unit in the mks system. To get yourself calibrated, one lbf is about 4.5 newtons; one newton is between $\frac{1}{4}$ and $\frac{1}{5}$ of a lbf, about the weight of a hamburger.

A third approach that is popular with American engineers (including the author) is the "engineering" system, in which standards for *both* force *and* mass are chosen arbitrarily. In this case, we cannot expect to find $F = Ma$, but we can expect to find $F = \text{constant} \times Ma$. So, to use this approach an engineer will write Newton's law as

$$F = \frac{1}{g_c} Ma \tag{2.9}$$

where g_c is a constant that depends on the arbitrarily chosen standards for force and mass. Equation (2.9) includes the mks and ft-lbf-slug systems discussed above, since for these cases $g_c = 1$. The engineering system uses a mass standard (you can think of it as the mass of a block of gold in some safe spot in Las Vegas) and a force standard (the weight of the block in Las Vegas). The weight of the standard block is called 1 lbf ("pound force"), and its mass is called 1 lbm ("pound mass"); thus, the weight in lbf of *any* object, as measured in or around Las Vegas, will be *numerically* equal to its mass in lbm. (Quick, what is your *mass* in lbm?) A car weighing 2,200 lbf will have a mass of 2,200 lbm. This makes it easy for engineers to determine the mass of something by weighing it,* which makes life easy for slow-witted engineers. The experimentally determined value of g_c in the engineering system of units is

$$g_c = 32.17 \text{ ft-lbm/lbf-sec}^2$$

To illustrate the use of Eq. (2.9), suppose we want to calculate the magnitude of the force that is required to accelerate the 2,200-lbm car to 60 mph in 10 sec. Now, 60 mph corresponds to 88 ft/sec; so the acceleration is

$$a = \text{velocity change/time} = \frac{88 \text{ ft/sec}}{10 \text{ sec}} = 8.8 \text{ ft/sec}^2$$

The force, which will be exerted on the car by the roadway at the rear tires, is then

$$F = \frac{1}{g_c} Ma = \frac{1 \text{ lbf-sec}^2}{32.2 \text{ ft-lbm}} \times 2,200 \text{ lbm} \times 8.8 \text{ ft/sec}^2$$

$$= 600 \text{ lbf}$$

As a second example, suppose we drop a 100-lbf weight; how fast will it accelerate as it falls? Quickly determining the mass as 100 lbm, we use Eq. (2.9) and find

$$a = g_c \frac{F}{M} = 32.2 \frac{\text{ft-lbm}}{\text{lbf-sec}^2} \times \frac{100 \text{ lbf}}{100 \text{ lbm}} = 32.2 \text{ ft/sec}^2$$

This is the "acceleration of gravity," usually denoted by g. Note that the car acceleration of 8.8 ft/sec^2 corresponds to about 0.27g. You might compute the force required to bring your car to rest from 60 mph in 0.1 sec, and then ask yourself if the vehicle and you would enjoy this adventure.

KINETIC ENERGY

When we accelerate an object, we increase its energy of motion, commonly called its "kinetic energy." Using the definition of work and Newton's law discussed above, an expression relating the kinetic energy to the speed of the object will now be derived. Consider an object being

---

* Within engineering accuracy this can be done at any point on earth, and one need not take the thing to Las Vegas.

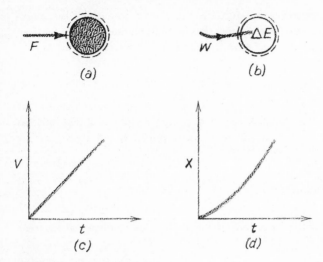

FIG. 2.15 SKETCHES USED IN DERIVING THE EXPRESSION FOR KINETIC ENERGY. (*a*) FORCES, (*b*) ENERGY FLOWS, (*c*) VELOCITY FOR CONSTANT *F*, (*d*) POSITION FOR CONSTANT *F*.

accelerated by a force; Fig. 2.15 defines the system and shows the energy flows. The energy balance on this system is

$$W \underset{\substack{\text{energy} \\ \text{input}}}{=} \underset{\substack{\text{increase in} \\ \text{energy storage}}}{\Delta E} \tag{2.10}$$

The increase in energy storage will be due to the increase in the kinetic energy; denoting the kinetic energy of the body by E_k, Eq. (2.10) becomes

$$W = \Delta E_k \tag{2.11}$$

If the body is initially motionless, it has no energy of motion, and $E_{k\,\text{initial}} = 0$. Thus, $E_{k\,\text{final}}$ represents the kinetic energy possessed by the body when all of W has been put in. If we can derive an expression for W, Eq. (2.11) will immediately give us an expression for the final kinetic energy E_k.

Let's take the special case where F is constant, in which case Eq. (2.9) tells us that a is also constant. If we start from rest, the velocity will increase linearly with time as indicated in Fig. 2.15c. The acceleration is V/t, where V is the velocity at a particular time t as shown in the figure. So, from Eq. (2.9), the force required to produce this acceleration is

$$F = \frac{1}{g_c} M \frac{V}{t} \tag{2.12}$$

Now, from Eq. (2.1), the work done by the force F in the time period it takes to reach V is

$$W = Fx \qquad (2.13)$$

where x is the distance traveled in this time. Now, the average velocity over this time period is $V/2$; so the distance traveled is

$$x = \tfrac{1}{2}Vt \qquad (2.14)$$

Combining the three equations above, we obtain

$$W = \frac{1}{g_c} (\tfrac{1}{2}MV^2) \qquad (2.15)$$

Finally, from the energy balance we see that this work is simply the kinetic energy that the body has when it is moving with velocity V; so

$$E_k = \frac{1}{g_c} (\tfrac{1}{2}MV^2) \qquad (2.16)$$

Equation (2.16) is the general expression for the "translational kinetic energy" of a moving body (if you have studied calculus, you should be able to show that this same expression holds irrespective of how F varies during the acceleration). You may have seen kinetic energy defined elsewhere as $\tfrac{1}{2}MV^2$; the g_c just gives us the flexibility of using a variety of unit systems (for the mks and cgs systems, $g_c = 1$ and $E_k = \tfrac{1}{2}MV^2$).

For example, let's calculate the kinetic energy of the 2,200-lbm car moving at 60 mph (88 ft/sec);

$$E_k = \frac{1}{32.2 \text{ ft-lbm/lbf-sec}^2} \times \frac{1}{2} \times 2{,}200 \text{ lbm} \times 88^2 \text{ ft}^2/\text{sec}^2$$

$$= 265{,}000 \text{ ft-lbf}$$

Neglecting friction, this much energy must be supplied by the motor to accelerate the car from rest to 60 mph.

As a second example of kinetic energy, let's estimate the kinetic energy of the wind flowing through a windmill. A good windmill would capture only part of this energy, and so this calculation will give us an upper limit on the energy that one could get from the wind. Figure 2.16 shows the system. Suppose that the air velocity is V. In a time interval t all the air in the column shown will pass through the windmill, and what we want to do is calculate the kinetic energy of this mass of air. A good work style is to first develop an expression in algebraic form and then at the end put in some numbers. We'll follow this format here. If we denote the

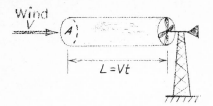

FIG. 2.16 CALCULATING THE POWER OUTPUT OF A WINDMILL.

windmill blade diameter by D, the cross-sectional area of the wind flow is

$$A = \frac{\pi D^2}{4} \tag{2.17}$$

Now, the length of the column is $L = Vt$; so the volume of the cylinder of air is

$$\text{Vol} = AL = \frac{\pi D^2}{4} Vt \tag{2.18}$$

To get the mass of the air, we need to know the *mass density* of the air; this is usually denoted by the Greek symbol ρ ("rho"), and for air at normal atmospheric conditions ρ is about 0.074 lbm/ft^3. That is, each cubic foot of air has a mass of 0.074 lbm. Now, the mass of the air column is

$$M = \rho \times \text{vol} = \frac{\pi D^2}{4} \rho Vt \tag{2.19}$$

The kinetic energy of this column is then, from Eq. (2.16),

$$E_k = \frac{1}{g_c} \left(\tfrac{1}{2} MV^2 \right) = \frac{\pi D^2 \rho V^3 t}{8 g_c} \tag{2.20}$$

To find the power, we simply divide the kinetic energy of the column by t, the time it takes for the air column to pass through the windmill; this is the *maximum* power that could be taken from the air by the mill,

$$\dot{W}_{\text{max}} = \frac{\pi D^2 \rho V^3}{8 g_c} \tag{2.21}$$

This is a neat little formula; it tells us that the available power varies as the *cube* of the wind velocity and as the *square* of the windmill diameter.

Let's put in some numbers. A typical mill might have a diameter of

20 ft, and a light breeze might be 4 ft/sec. We'll use the air-density value given earlier. Then,

$$\dot{W}_{max} = \frac{\pi \times 20^2 \text{ ft}^2}{8 \times 32.2 \text{ ft-lbm/lbf-sec}^2} \times 0.074 \text{ lbm/ft}^3 \times 4^3 \text{ ft}^3/\text{sec}^3$$

$$\dot{W}_{max} = 23 \text{ ft-lbf/sec}$$

Noting that one horsepower is 550 ft-lbf/sec, we see that the *maximum* power available from the windmill is only about $23/550 = 0.04$ hp! However, suppose the wind is four times faster, at 16 ft/sec (rather brisk). Then, the maximum power increases by a factor of 4^3 to about 2.5 hp. It will still take a lot of windmills to run a farm!

No real windmill could deliver the maximum power. The *efficiency* of the windmill would be defined as the ratio of the actual power to the theoretical maximum; the efficiency is usually denoted by the Greek symbol η ("eta"), and thus

$$\eta = \frac{\dot{W}_{actual}}{\dot{W}_{max}} \tag{2.22}$$

where $\dot{W}_{max}$ is given by Eq. (2.21). A real windmill might have an efficiency of only 20%. There has not been much research to develop better windmills simply because the available energy is so very small.

To summarize, kinetic energy is a mechanical form of energy associated with the motion of mass; its value may be computed from Eq. (2.16).

POTENTIAL ENERGY

Consider the weight being hoisted by the cable in Fig. 2.17. As the weight is lifted, energy is transferred *to* the weight as work *from* the cable by the cable force F_c. Also, energy is transferred *from* the weight *to* the earth's gravitational field by the action of the weight force F_g. The energy balance on the weight system is

$$\underset{\substack{\text{energy} \\ \text{in}}}{W_c} = \underset{\substack{\text{energy} \\ \text{out}}}{W_g} \tag{2.23}$$

where W_c and W_g denote the energy transfers as work just described (note that we presume that the motion is slow, so that the kinetic energy is not important). Now, the work W_g is

$$W_g = F_g h \tag{2.24}$$

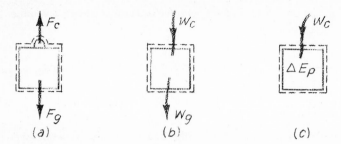

FIG. 2.17 THE WORK DONE ON
THE EARTH'S GRAVITATIONAL FIELD
IS COMMONLY THOUGHT OF AS
AN INCREASE IN THE POTENTIAL
ENERGY OF THE WEIGHT.
(*a*) FORCE DIAGRAM. (*b*) ONE WAY
TO THINK OF ENERGY FLOWS.
(*c*) A MORE COMMON WAY TO
THINK OF ENERGY FLOWS.

where h is the height that the weight has been raised. So Eq. (2.23) is

$$\underset{\substack{\text{energy} \\ \text{in}}}{W_c} = \underset{\substack{\text{energy} \\ \text{out}}}{F_g h} \qquad (2.25)$$

In this interpretation of the energy balance there is no change in the energy of the weight. An alternative view in common use is based on a reinterpretation of Eq. (2.25); since h is zero initially, the above equation is equivalent to

$$\underset{\substack{\text{energy} \\ \text{in}}}{W_c} = \underset{\substack{\text{change in energy} \\ \text{storage}}}{\Delta(F_g h)} \qquad (2.26)$$

The term on the right in Eq. (2.26) is now interpreted as a change in the energy stored within the weight, rather than an energy outflow to the gravitational field as in Eq. (2.25). The "potential energy" of weight is defined as

$$E_p = wh \qquad (2.27)$$

where w is the weight force and h is the height that the weight is raised above the ground. When using the potential-energy concept, energy transfers as work by gravitational forces are not put into the energy balance as work terms, but appear instead as changes in the potential energy (see Fig. 2.17c).

To illustrate the application of potential energy, consider a mass that is dropped from a height h above the floor. Neglecting wind friction, there

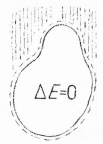

FIG. 2.18 ENERGY FLOWS FOR A FREELY FALLING OBJECT.

will be no energy transfers to or from the body during the free-fall period. So an energy balance on the mass gives (see Fig. 2.18)

$$
\underset{\substack{\text{energy} \\ \text{in}}}{0} = \underset{\substack{\text{energy} \\ \text{out}}}{0} + \underset{\substack{\text{change in} \\ \text{energy storage}}}{\Delta E} \tag{2.28}
$$

where E represents the energy of the mass, i.e., the sum of its potential and kinetic energies,

$$
E = E_k + E_p = \frac{MV^2}{2g_c} + wh \tag{2.29}
$$

So the energy balance simply says that the final and initial total energies are the same. If the body is initially at rest with zero kinetic energy, and if $h = 0$ at the end of the drop so that the final potential energy is zero, we have

$$
E_{k \text{ final}} = E_{p \text{ initial}} \tag{2.30}
$$

So

$$
\frac{1}{2g_c} MV^2_{\text{final}} = wh_{\text{initial}} \tag{2.31}
$$

The relationship between the weight and mass is given by Newton's law [Eq. (2.9)] as (recall that the free-fall acceleration is $g = 32.2 \text{ ft/sec}^2$)

$$
w = \frac{1}{g_c} Mg \tag{2.32}
$$

Using this in Eq. (2.31) for w, and solving for V_{final}

$$V_{\text{final}} = \sqrt{2gh_{\text{initial}}} \qquad\qquad (2.33)$$

Note that it is g and not g_c that appears in this expression, and that the final velocity is independent of the mass of the body (remember Galileo's fable about the feathers and the stones?).

As a second example, let's calculate the potential energy of rain in a cloud. Suppose the clouds at an elevation of 2,000 ft contain enough water to dump 1 in of rain on the ground below. For each square mile of ground we will have $(5,280)^2 \times (\frac{1}{12}) = 2.3 \times 10^6$ cubic feet of water. The *mass density* of water is about 62.4 lbm/ft^3; so the *weight density* of water is 62.4 lbf/ft^3 (do you see the subtle but important difference here?). Using the definition [Eq. (2.27)], the potential energy of this water mass is

$$E_p = wh = 2.3 \times 10^6 \text{ ft}^3 \times 62.4 \text{ lbf/ft}^3 \times 2{,}000 \text{ ft}$$

$$= 2.9 \times 10^{11} \text{ ft-lbf}$$

This is a lot of energy; the sun had to supply at least this much energy to raise the water to the 2,000-ft level through atmospheric circulations. Now, if this water falls in a deluge over a period of 100 min, the rate at which potential energy is being released corresponds to a power of

$$\dot{W} = 2.9 \times 10^{11} \text{ ft-lbf}/10^2 \text{ min} = 2.9 \times 10^9 \text{ ft-lbf/min}$$

$$= 87{,}000 \text{ horsepower}$$

Remember that this is for 1 square mile. Perhaps someday clever engineers will find ways to capture some of this energy, and rainpower will be one of the ways that man utilizes the sun's energy more directly!

Man's use of nature's elevation of water has so far been confined to hydroelectric power, in which the potential energy of water in the mountains is converted to electrical energy in hydraulic turbines. As a final calculation example, let's estimate the hydroelectric power derivable from a mountain river. Suppose the river is 50 ft wide, 20 ft deep, and has an average flow velocity of 2 ft/sec. Suppose we plan to build a dam to raise the elevation of the reservoir water 200 ft above the hydroelectric plant. This is enough information to estimate the maximum power that could be generated from this energy source. Figure 2.19 shows the system that we analyze. We will compute the potential energy in the column of water shown on the figure; this is the amount of water that will enter

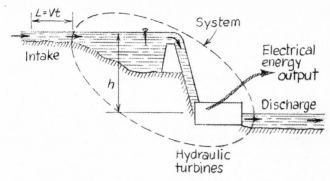

FIG. 2.19 ESTIMATING THE POWER OUTPUT AVAILABLE FROM THE POTENTIAL ENERGY OF A RIVER FLOW.

the reservoir in time t, which for steady conditions will be the amount of water that is passed through the hydraulic turbines. Then we will divide by t to get the power as in the windmill analysis. The volume of the water column is

$$\text{Vol} = AVt \qquad (2.34)$$

where A is the cross-sectional area of the river. Denoting the weight density (lbf/ft^3) by the Greek symbol γ ("gamma"), the weight of the column is

$$w = AV\gamma t \qquad (2.35)$$

Now denote the height that we can drop the water by h. Then the potential energy that we have available is

$$E_p = hAV\gamma t \qquad (2.36)$$

and the power available is

$$\dot{W} = AV\gamma h \qquad (2.37)$$

You might note that the product AV is the "volume flow rate" in the river (ft^3/min) and the product $AV\gamma$ is the "weight flow rate" (lbf/min). Often you can find data in this form for rivers in various areas.

Let's put in the numbers:

$$A = 50 \text{ ft} \times 20 \text{ ft} = 1,000 \text{ ft}^2$$

$$\dot{W} = 1,000 \text{ ft}^2 \times 2 \text{ ft/sec} \times 62.4 \text{ lbf/ft}^3 \times 200 \text{ ft}$$

$$= 2.5 \times 10^7 \text{ ft-lbf/sec}$$

We can express this in some other power units (you should check the conversions if you don't do them easily),

$$\dot{W} = 2.5 \times 10^7 \text{ ft-lbf/sec} = 45,000 \text{ hp} = 34,000 \text{ kw}$$

So the available power from this station would be only 34 megawatts, which you can compare with the 1,000-Mw outputs of large fossil-fuel power stations. Perhaps now you can see why hydroelectric power is relatively unimportant in the United States; there is just not much hydroelectric energy available.

The "efficiency" of a hydroelectric power plant is the ratio of the *actual power produced to the theoretical maximum given by Eq. (2.37)*. Modern hydroelectric plants have efficiencies of the order of 60%; so we might realistically expect to generate only 0.6 × 34 = 20 megawatts on our river. What happens to the rest of the available energy is something that we will learn more about in the next chapter.

The preliminary engineering analysis for a proposed hydroelectric plant will follow exactly the calculation we just made. First the maximum available power will be calculated from the flow and elevation data, and then the actual power will be computed using an efficiency based on experience. There is no magic to engineering calculations, just a little science and lots of common sense!

TURBINES, PUMPS, AND COMPRESSORS

Let's see how these ideas are incorporated in modern energy-conversion technology. One of the earliest machines for producing useful mechanical energy was the hydraulic turbine. In its earliest and simplest form, water from a small stream was dumped onto paddles attached to a rotating wheel. The wheel turned a shaft and the shaft turned a grindstone or other tool requiring the input of energy as work. This waterwheel has very simple energetics; the potential energy of the water is converted to kinetic energy during the fall. The impact of the moving water on the paddles produces a downward force to turn the wheel, and the motion of this force gives energy to the wheel as work. Perhaps 10 to 20% of the potential energy available in the water flow can be converted to work in simple devices of this type. Additional potential energy can be recovered in the wheel if the paddles are made more bucketlike. The advance of waterwheel technology is indicated by the modern Pelton wheel shown in Fig. 2.20. Water taken from below the surface of a reservoir, or from another high-pressure source, is fed through a firehoselike nozzle, from which it emerges at a very high velocity (perhaps 300 ft/sec). This jet impacts upon the buckets, is turned, and is discharged with a much lower velocity relative to the nozzle. There is a reduction on the kinetic energy of the stream as a result of this lowering of its speed, and this is accounted for by work that has been delivered to the buckets. Pressure forces on the fluid are required to turn it and to slow it down; these pressure forces are provided by the bucket, and it is these forces that push the buckets to turn the wheel around. A modern Pelton wheel would have a maximum efficiency of conversion of potential energy to useful work of about 80%.

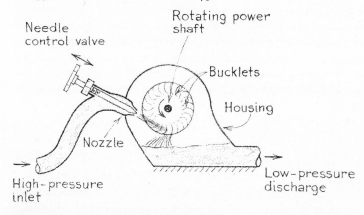

FIG. 2.20 PELTON-WHEEL SCHEMATIC.

In "impulse turbines," such as the Pelton wheel, the fluid stream is accelerated in a nozzle, but the flow velocity *relative to the bucket* does not vary as the flow passes through the bucket. Another type of turbine device is the "reaction turbine." Reaction turbines operate completely filled with fluid and with "blades" on the rotor rather than buckets. The area for flow decreases as the flow moves between the blades, and hence the fluid is also accelerated with respect to the rotor within the blade passages. This alters the pressures exerted by the fluid on the blades, and the design is such that these pressures help push the rotor around.

Let's think for a moment about the performance characteristics of an

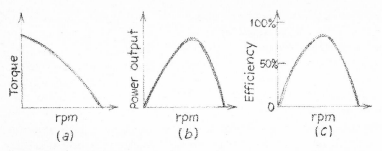

FIG. 2.21 PERFORMANCE
CHARACTERISTICS OF AN IMPULSE
HYDROTURBINE. (*a*) TORQUE,
(*b*) OUTPUT POWER, (*c*) EFFICIENCY
OF CONVERSION OF FLUID ENERGY
TO SHAFT WORK.

impulse hydraulic turbine. The jet of water will squirt onto the buckets, regardless of the bucket speed. The greatest force on the buckets will occur when the wheel is prevented from rotating, but as the wheel speed increases the force on the buckets will decrease. Thus, the torque-speed* characteristics of the machine will look something like Fig. 2.21a. Note that at a certain speed the torque will fall to *zero*; this is the speed at which the buckets and the water really don't interact, for the buckets are moving so fast that they just run with the water. Now, force times distance is work; so force times velocity is power. You can derive for yourself the relationship

$$W = 2\pi TN \tag{2.38}$$

where W is the power output, T the torque, and N the number of revolutions per minute of the wheel. Working from Fig. 2.21a, the power will be a curve like Fig. 2.21b. Note that it will peak out at about half the runaway speed, or at about half the water-jet speed. Hence, there is a wheel speed at which the maximum power can be obtained. Since this curve is drawn for a fixed jet speed and flow (i.e., for a fixed available kinetic energy) the efficiency curve will be the same shape as the power curve, and peak efficiency will also occur at the peak power point. The rpm at this condition would be called the "design speed" of the machine.

One always tries to operate a turbomachine at its design speed. The power output can be controlled by adjusting the *amount* of water flow. If too little power is demanded by the turbine load, the wheel will start to speed up; if the load draws too much power, the wheel will start to

---

* Torque is the circumferential force times the radius to the force.

slow down. In hydroelectric power stations, controls are used to sense the turbine speed and then to alter the water flow to hold the speed constant, automatically adjusting the output power.

Steam turbines operate on much the same idea. High-pressure steam is fed to nozzles, where it is accelerated, and then this high-speed flow (perhaps at 2,000 ft/sec) impacts upon turbine blades attached to a rotor. The flow is turned and slowed down relative to the ground, and the pressure forces to do this job are provided by the blades. The equal-but-opposite forces on the blades push the rotor around, producing useful power. Small steam turbines admit steam only to a segment of the rotor, as in the Pelton waterwheel, while large turbines operate with "full admission" of steam all around the perimeter. Most steam turbines are a mix of impulse and reaction types, with the design being tailored to achieve maximum efficiency for the application at hand. Figure 2.22 shows a large steam turbine; as the steam passes through the machine and its pressure drops, the steam expands to take up more space. In order to keep about the same flow velocities in the various stages of the turbine, the low-pressure stages must have larger flow areas, as is clearly evident in the picture. Steam turbines are usually "axial-flow" machines, where the basic direction of flow is along the axis of rotation. The nozzles accelerate the flow to high velocity and introduce strong swirl, and the idea is to capture as much as possible of the kinetic energy of the flow by removing the swirl in the turbine-blade passages attached to the rotor.

The "efficiency" of the turbine describes the fraction of the available energy that is captured as shaft work.* Large steam turbines have peak efficiencies of the order of 85%. In smaller turbines such as might be of interest for automotive application, leakages, friction, and other effects reduce the performance, and a machine in the range of a few tens of horsepower might have an efficiency of only 60%.

Gas turbines using air, helium, or other gas as the "working fluid" are of increasing importance in energy technology. They are a key component in gas-turbine engines, are well suited for power generation in the range of tens of kw to a few Mw, and thus have growing industrial applications. A significant number of trucks and busses are now powered by gas-turbine engines, whose air-pollution characteristics and light weight make them well suited for transportation uses. The performance curves for gas turbines are similar to Fig. 2.21. Large gas turbines have peak efficiencies of the order of 85%, while a machine in the power range for automotive application might have an efficiency of around 80%.

---

* The efficiency of energy conversion in the turbine itself should not be confused with the overall energy conversion of a gas-turbine *engine*, which would be 25 to 35%.

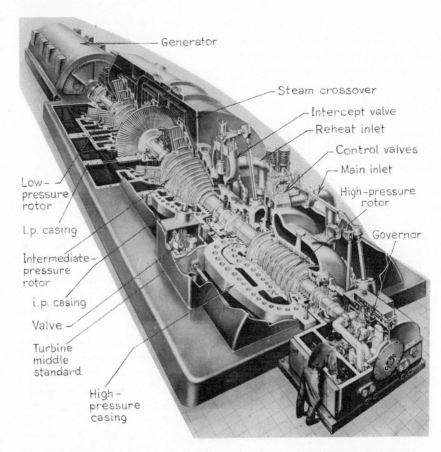

Generator

Steam crossover

Intercept valve

Reheat inlet

Control valves

Main inlet

High-pressure rotor

Governor

Low-pressure rotor

l.p. casing

Intermediate-pressure rotor

i.p. casing

Valve

Turbine middle standard

High-pressure casing

FIG. 2.22 LARGE STEAM-TURBINE CUTAWAY. COURTESY OF THE GENERAL ELECTRIC COMPANY.

Gas-turbine engines have other turbomachinery components. The engine requires high-pressure gas, which is obtained by compressing the intake flow. Aircraft engines normally employ "axial-flow" compressors, which look something like axial-flow turbines operating in reverse. The flow enters the compressor without swirl, and high-velocity swirl is imparted to the flow by the rapidly rotating rotor blades. The kinetic energy of this swirl is removed by a row of stationary blades, resulting in an increase in the gas pressure. This process is repeated through several rotor and "stator" stages until the desired pressure increase has been obtained.

The "fan" stage of a modern aircraft engine is nothing but a second axial-flow compressor driven by the main engine. The fan compresses air to a high pressure, and then this air is blown out through a nozzle to produce the thrusting force. When you visit an airport, look at the engines carefully; the earlier engines had no fans, and the first fan engines ran only a small amount of flow through the bypass fan. The newer engines, such as those used on the Boeing 747 and DC10, have large fans that pump about five times as much flow as is handled by the main engine, and their fans are quite noticeable at the front end of the engines.

Another device that is used for some lower-power applications is the "radial-flow" or "centrifugal-flow" compressor. The flow enters near the axis of rotation and is hurled outward by the rapidly spinning blades. Work done by the blades therefore appears as kinetic energy in the moving gas. The high-speed flow leaving the rotor is then slowed down by deceleration in a "diffuser," and this produces an increase in the gas pressure. Centrifugal compressors are used for supercharging of piston engines, blowing air into furnaces, and in automotive gas-turbine engines. Compressor efficiencies compare the actual machines with hypothetical devices in which all the energy input is put into a fully available form. Efficiencies of 80% are typical for large multistage units, while smaller machines used for blowing air have efficiencies closer to 50%.

A centrifugal pump, such as used to circulate coolant in automobile engines, works on the same idea. Liquid enters on the axis and is thrown outward by the rotating blades; this is the process that increases the energy of the liquid flow. The kinetic energy of the liquid is converted to "flow energy" (a term we shall discuss in Chap. 4) by the deceleration of the flow in the diffuser. This produces the high discharge pressure at the pump outlet. The performance characteristics of a centrifugal pump or compressor depend on the blade design, and one typical example is shown in Fig. 2.23. The highest discharge pressure for such a pump occurs when the flow is shut off and the impeller spins in stationary fluid. As the discharge valve is opened to permit flow, the discharge pressure drops, producing the typical pump characteristic curve shown in Fig. 2.24. The pump becomes ineffective at high flow rates, and hence there is a flow at which the efficiency is a maximum. For large hydraulic units the peak efficiency might be of the order of 80%, while for smaller units capable of pumping a few hundred gallons per minute the efficiency might only be about 50%.

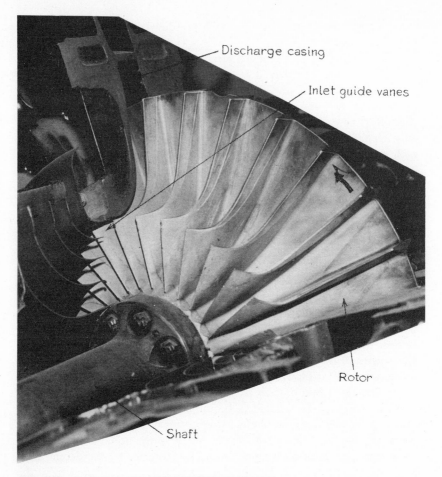

Discharge casing

Inlet guide vanes

Rotor

Shaft

FIG. 2.23 RADIAL FLOW COMPRESSOR CUTAWAY.

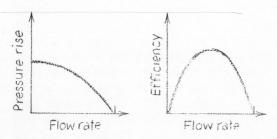

FIG. 2.24 CHARACTERISTICS OF A CENTRIFUGAL PUMP.

PROBLEMS

Developing skill with numbers

2.1 A car with a 200-hp engine running and in gear but with clutch slipping pushes hard against a stationary wall. How much power is the car delivering to the wall?

2.2 The steel used in a cable has a yield stress of 200,000 lbf/in². Safe practice might allow working stresses of half this magnitude; calculate the safe load that can be lifted with a cable having a diameter of $\frac{1}{4}$ in.

2.3 The footings for a tower will be placed on earth that can support a compressive load of 2,000 lbf/ft². Calculate the foundation area required if the tower weight is 80,000 lbf.

2.4 A hydraulic system can supply oil at a pressure of 400 lbf/in². This is to be installed on a piece of farm equipment to operate hydraulic-load cylinders of 3 in diameter. Calculate the force that can be developed with each of these cylinders.

2.5 For the cable of Prob. 2.2, what will happen if the load is accelerated upward at $2g$? What is the load that can be accelerated upward safely at $2g$?

2.6 A man lifts a 150-lbf barbell up 2 ft. How much work (ft-lbf) did he do? Suppose he does this 20 times in 1 min; how much power does he expend in this exercise? When he lowers the weight, does he recover any energy useful for lifting weights? What happens to the energy he does recover? How does the energy he uses in 20 presses compare with the food energy he takes in in 1 day?

2.7 A climber wants to develop a pneumatic device to shoot ropes uphill. Using air at 40 psia from a small pressure bottle, he wants to be able to shoot a 20-lbf weight up a height of 100 ft. He has built a simple piston launcher with a 1-in-diameter piston that can travel in the launch tube a distance of 10 in. He plans to attach the weight to the piston, which itself weighs 1 lbf, pressurize the launcher, and then release the piston to hurl the weight and piston upward. Assuming that the full air pressure acts on the piston for the full 10-in distance, calculate the height that his device would toss the weight. If it will work, how much excess capacity does it have? If it will not work, how should the design be changed so that it will?

2.8 Calculate the increase in potential energy of a 2,500-lbf automobile that has climbed from sea level to 5,000 ft. How did the car get this energy?

2.9 Assume that the car in Prob. 2.8 can coast back to sea level without friction or wind resistance, with no brakes; calculate the velocity that the car will have when it reaches sea level.

2.10   A 40,000-lbf truck descends from 4,500-ft elevation on a winding mountain road to 3,700-ft elevation. Calculate the potential energy decrease of the truck, and explain what happens to this energy, assuming that the driver maintains a uniform speed.

Some more interesting analyses

2.11   Perform a series of experiments on yourself to determine your capabilities as an energy and power producer (lift weights, etc.). Estimate the amount of energy that you could produce in 1 day working at a reasonable level of effort, and the average power for the 8-hr work period. Estimate the power that you deliver over a shorter time period. Calculate your efficiency as a machine, taking your caloric intake as the energy-input term and your 8-hr output as the energy-output term. Estimate the cost of your food energy, and calculate the cost of your 8-hr energy output on a cents/kwhr basis. Compare this with the going rate from electrical companies of about 2.5 cents/kwhr. Decide what you would ask to be paid to work like this for an 8-hr day, and then calculate the value of your energy output on a cents/kwhr basis. Estimate the amount of energy you use or control in 1 day for lights, space heating, transportation (0.5 hp-hr/passenger-mile is a good figure for automobile transportation), etc. Calculate how many hours you would have to work to produce this much energy yourself; calculate how much you would charge for this energy at the cost for your labor determined above.
Discuss the importance of energy in your life and how these calculations have affected your values in regard to energy.

2.12   It has been proposed that a low-level dam be built at the north end of San Francisco Bay in the Carquinez Strait. The dam would be used primarily for salt-water-intrusion control, and the lake behind the dam would provide recreational facilities for the area. The proposed dam would increase the water level at the damsite by 15 ft, and power would be generated by dropping the river flow through hydraulic turbines to the bay below. The river-volume-flow rate is reported to be 30 million acre-feet per year. Calculate the power capability of this hydroelectric station, and the number of kwhr of energy that could be generated by this station each year. Would this handle the annual demand of the BART (Bay Area Rapid Transit) system of 320,000,000 kwhr/year?
Familiarize yourself with the geography, demography, and ecology of the area, and discuss the values questions that relate to debate on this proposal.

2.13   It has been proposed that a tidal power station be built across the Golden Gate to San Francisco Bay. Large hydraulic turbines would generate power from the tidal flow in and out of the bay. The area of the bay is about 1,200 square miles, and the difference between high and low tide is on average about 4 ft. Assuming that the same volume of flow will occur through the power station, and that the average drop of water will be lowered 2 ft as it passes through the power station, calculate the maximum power that could possibly be developed with such an installation and the "actual" power, using an estimated efficiency of 30%. Compute the energy (kwhr) that could be developed in 1 year, and compare this with the requirements of the BART system (see Prob. 2.12).

Discuss other technical and nontechnical questions that should be answered in the consideration of this proposal, and the way in which the values of the local society might affect the final decision.

2.14 The use of electrical energy varies during the day by the order of 30 to 50%. Since there are no good ways to store large amounts of electrical energy, power plants adjust their output during the day to meet the demands of the moment. This means that many plants are operating at part load some of the time, and at part load the efficiency is almost always less than at full load. Moreover, the load patterns mean that the installed power capacity must be about twice as large as the maximum demand. For these reasons, utility companies are very interested in schemes that would permit them to generate electrical energy at a uniform power throughout the day, storing this energy somehow for use during periods of peak demand. Hydroelectric stations already are an important part of this system, for they can shut down completely during low-load periods, allowing the fossil-fuel power stations to keep running at near-peak efficiency. A scheme that is attracting current interest involves using the excess electrical energy produced during low-load periods to pump water to uphill reservoirs, and then to let this water run downhill through a hydroelectric power station to generate more power for the peak-load situations.

To study this approach, consider a hybrid power station consisting of a 400-Mw fossil-fuel power plant and a 100-Mw hydroelectric plant. Suppose that a 400-ft elevation difference exists between the two reservoirs of the hydroelectric station. Calculate the volume flow of water through the hydroelectric plant when it is generating its full capacity of 100 Mw (assume a generating efficiency of 70%). Assume that the combined system delivers full power for 8 hr, and operates at a uniform off-peak power for 16 hr each day. Then, calculate the power that is required to return this flow to the upper reservoir, assuming that the pumping efficiency is 60% (only 60% of the electrical energy input to the pump goes to moving the water uphill). What is the net power output from the combined plant when both units are generating power, and what is the net power output when water is being pumped uphill? Assuming that recreational uses of the reservoirs require that the water level not change by more than 2 ft over the day, calculate the acreage that must be allotted for each reservoir.

Choose an area familiar to you where a system of this sort might be installed. Discuss the geographical, ecological, and demographic aspects of the area that would be important in the decision on whether or not to build this system there, and the private and public values that might come into conflict during debate on this question.

2.15   This problem deals with the energy requirements for a rapid-transit system. Choose a major city near you that does not at present have a rapid-transit system. On a map of the area, lay out a network of tracks and stations that would serve the main bedroom communities around the city, say for a radius of 30 miles from the heart of town. An initial system might have of the order of 20 to 40 stations; so you can place these at strategic locations in the suburban and downtown areas. This initial rough plan will give you some feeling for the rail lengths of the transit system. Next, decide upon the approximate service frequency at each station, and the number of people that will commute along each line. This should provide you with enough information to estimate the number of trains your system will need; assume that the trains are independently powered and each can carry 80 passengers at a maximum speed of 80 mph. For comparison, the BART system in the San Francisco Bay Area will use about 450 72-passenger vehicles.

The first calculation involves the installed power capacity in each vehicle. The primary factors determining this power are the acceleration capability and the grade-climbing capability. Assume that your cars weigh 60,000 lbf each empty (the smaller BART cars weigh 56,000 lbf). You probably want to be able to climb a 3% grade at 80 mph (up 3 ft for each 100 ft traveled), and this requirement fixes the "grade-climbing power." As you leave a station, you want to be able to accelerate on level tract to 80 mph, and an acceleration of $0.1g$ seems tolerable. This information will allow you to calculate the acceleration power. The motor for each vehicle should deliver enough power to accommodate the larger of these two requirements; aerodynamic power and rolling friction are smaller. For comparison, the BART cars have 600-hp capability in each vehicle.

Next, estimate the maximum power that your system will demand. Some cars will be accelerating while others are cruising, so you aren't likely ever to call for the full 600 hp for every vehicle. Your estimated power is at best an educated guess at this point, but could be refined by very detailed analysis of the operation of the system.

The previous calculation gave you an estimate for the maximum power that you would request from the electric company. You will normally operate at a lower power level; estimate your average power level, and then calculate the annual energy requirement of your system in kwhr. For comparison, the BART system will require around 320,000,000 kwhr/year.

Individual commuting by automobiles requires about 0.5 hp-hr of energy per passenger-mile. Compare the energy cost of personal commuting with that of your rapid-transit system. Suppose that the electric company charges your transit company 2.4 cents/kwhr for electrical energy; what is the approximate energy cost on a cents-per-passenger-mile basis, and what will be the magnitude of the transit company's annual electrical bill?

Discuss the economic, political, demographic, and ecologic factors that would be involved in the development of a rapid-transit system for this area, and the manner in which the values of individuals and the community would influence the decision on the construction of such a system.

Some research problems

2.16 "Hydroelectric power does not require or consume fuels; so in an under-developed country a little bit of hydroelectric power can have a large impact." Investigate the history of energy development in emerging countries, and write a critique of this statement.

2.17 Choose an underdeveloped country. Familiarize yourself with its geography, rainfall, and runoff patterns, and identify the likely locations for new hydroelectric power stations. Estimate the power that could be produced in these stations, and examine the impact of this new energy on the development of the country.

Discuss the values of that society as they pertain to the development of additional electrical energy within the country, concentrating in particular on the differences between these values and those that you see operative in the United States.

2.18 Trace the development of hydroelectrical power in the United States, and write a critical analysis of the statement "the value system in the United States works against the development of hydroelectric power."

2.19 Read up on the controversy surrounding the Glen Canyon Dam project on the Colorado River. Identify the technical and nontechnical factors that influenced this debate and the various values that different groups felt were important in relation to the issue. Write a review of what happened and a critique of this from the background of your own values.

2.20 Trace the development of windmill technology, and write a short review of your studies. Be quantitative wherever possible. In what contemporary cultures is windpower still important?

3

HEAT AND
OTHER THINGS

In which the order
and chaos of molecular motion
are assigned their due

INTERNAL ENERGY, HEAT, AND TEMPERATURE

We have said that all matter has energy—but how much? Very shortly we will see some equations, charts, and tables that tell us the amount of energy in various substances at various temperatures and pressures. But first we need to understand *how* matter has energy, and to think very carefully about the ways that the energy stored in a glob of some substance can be changed.

There are two basic approaches to studying the energy of matter. In the *microscopic* approach one thinks in terms of matter at the atomic and molecular level. This is very useful for understanding the behavior of matter, and for calculating some of the properties of matter from basic physics. But when one goes to analyze practical systems, which might contain from 10^4 to 10^{30} molecules, it is impractical to deal with each molecule, and another approach is taken. This is the *macroscopic* approach, in which matter is treated as being *continuous*. Practical analysis of energy systems is almost always carried out using the *macroscopic* approach. However, *microscopic* models are very useful for understanding *macroscopic* analysis, and so one usually thinks about what is happening to molecules even when doing macroscopic analysis.

Let's review a model of the microscopic nature of matter that helps explain and understand energy. In this model, matter is thought of as being made up of molecules, which in turn are made up of atoms, which in turn are made up of electrons, protons, and neutrons, which in turn are made up of . . . well, that's what physicists are trying to decide today. Each of these particles has a tiny amount of kinetic energy associated with its motion and potential energy associated with forces between the particles. The sum of all the energies of all the microscopic particles in a glob of matter is called the "internal energy" of the glob in the macroscopic approach to energy.

In a solid at low temperature the main contributions to internal energy arise from strong potential energies associated with the forces that hold the molecules close together, and from kinetic energy associated with jittering motions of the molecules; the higher the temperature, the more energetic the vibrational jitter. Melting occurs when the vibration becomes vigorous enough to break these force bonds, allowing the molecules to slip freely around one another. Evaporation occurs when the molecules are given enough kinetic energy to escape the strong attraction of their neighbors, and fly away to form a gas. In a gas the primary contribution to internal energy comes from the translational kinetic energy of the molecules. The molecules may also rotate and vibrate, especially at higher temperatures, and this too contributes to the internal energy. At very high temperatures electrons are knocked out of the molecules by collisions with other molecules. Through this process of "ionization" the gas becomes a "plasma"; the free electrons in a plasma make it a good conductor of electricity. You probably have seen plasmas in action in the atmosphere and in some types of lamps; these ultra-high-energy fluids will be very important in the energy technology of the future.

We can't get our hands on individual molecules; so how can we measure the internal energy? One way to study the internal energy of a particular substance macroscopically is to transfer some energy to the substance as work; a measurement of the work then determines the *change* in the internal energy through the energy balance. For example, suppose we compress a gas in the piston-cylinder system shown in Fig. 3.1.

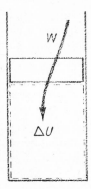

FIG. 3.1 AS THE PISTON PUSHES THE GAS IN, THE INTERNAL ENERGY OF THE GAS INCREASES.

As the piston compresses the gas, it does work; W can easily be computed from measurement of the force exerted and the distance moved. Then, if we denote the internal energy of the gas by U, as is customary, the energy balance is

$$\underset{\substack{\text{energy}\\\text{input}}}{W} = \underset{\substack{\text{increase in}\\\text{energy storage}}}{\Delta U} \tag{3.1}$$

Equation (3.1) permits the final internal energy to be calculated if the initial internal energy and energy transfer as work are known. The internal energy is usually chosen to be zero at some arbitrary condition; so experiments like this would permit the evaluation of the internal energy of the substance under other conditions.

A microscopic model of this work process will be helpful. When the piston compresses the gas, the molecules of the piston march slowly ahead in a stately parade. Gas molecules continually bombard the piston (this is what makes pressure), and the slow, steady advance of the piston gives a little extra push to each gas molecule that bounces off it. This increases the energy of the gas molecules, that is, the internal energy of the glob of gas.

In the macroscopic approach, work is not the only way to change the internal energy. Think of a gas in a metal container (Fig. 3.2). If the

FIG. 3.2 HOW ENERGY IS TRANSFERRED AS HEAT.

container is very hot, its molecules will be vibrating furiously. If the gas is relatively cold, its molecules will be moving about sluggishly, and whenever one strikes the wall the metal molecules will kick it away smartly, thereby increasing the kinetic energy of the gas molecule and reducing the vibrational energy of the metal molecules. Many such collisions will result in substantial increase in the internal energy of the gas, and a reduction of the internal energy of the container. At the microscopic level this energy-transfer process is clearly work. But in a macroscopic analysis it is not counted as work because no macroscopically measurable forces move matter in a macroscopically observable manner. So in the macroscopic approach this energy transfer is called "heat" or, better yet, "energy transfer as heat." Heat is a microscopic transfer of energy as work that can't be measured by measuring macroscopic forces and distances (we would have to measure the work done by each and every solid molecule). As we shall discuss shortly, the energy transfer as heat will take place from the material at the higher temperature to the material at the lower temperature.

Is it reasonable to think of hot matter as having lots of heat? Imagine shrinking and going into a cloud of molecules whizzing about in all directions. You might be able to measure the total energy, but you could not tell if they got this energy by being squeezed in by a piston or by being kicked smartly by a quivering chaos of excited wall molecules. In other words, once the energy had found its way into the gas, you could not tell if it got there as heat or as work. Therefore, one should not think about heat as energy stored in matter or divide the internal energy of matter into a heat part and a work part. Early scientists tried to make such a distinction (some fuzzy thinkers still do), and great confusion arose around the meaning of the term "heat." It is *incorrect* to use heat to mean all or part of the internal energy, but some people still do; so be prepared to translate improper scientificese. "Heat" and "work" are both terms properly used only to describe energy *transfer*, observed at the boundaries of a system, and should never be used to describe energy storage.

Just as we can transfer energy in or out of a system as work, we can transfer energy in or out as heat. To put energy in as heat, we need to bring a hotter thing into contact with the system, and to take out energy as heat, we need to use a colder thing. In a household refrigerator things get cold because there is colder "freon" (a fluid used in refrigerators) inside the cooling coils. The energy removed from the meat is in turn put into the kitchen air from hot freon flowing in the coils behind the refrigerator (feel the warmth on yours). A hot-water bag is used to put energy into a cold foot; an ice bag is used to take energy out of a hot head. The science and technology of "heat transfer" is a very important subculture of the science and technology of energy, and high-technology heat transfer is making some remarkable things possible, such as drastic reduction in the size of heat-exchange equipment and the ability to transport energy as heat over great distances efficiently.

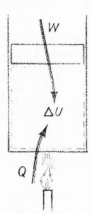

FIG. 3.3 HEATING AND WORKING TOGETHER.

Suppose we have a system into which we put some energy as work and some more as heat (Fig. 3.3). The energy balance on this system is

$$\underset{\substack{\text{energy} \\ \text{input}}}{W \; + \; Q} \; = \; \underset{\substack{\text{increase in} \\ \text{energy storage}}}{\Delta U} \tag{3.2}$$

If we can measure W and Q, we can compute ΔU; or if we can measure W and ΔU, we can compute Q. The energy balance can always be used to calculate one of the terms in it, provided all the other terms are known.

Temperature is a macroscopic concept that enters the picture at this point. You no doubt have some familiarity with temperature but probably have never thought very scientifically about the concept. Temperature is one of the most sophisticated concepts in macroscopic science. It is intimately tied up with the concept of energy transfer as heat just discussed. Energy transfer as heat always takes place *from* a body at higher temperature *to* a body at a lower temperature; heat flows "down the temperature hill." This is not experimental fact but rather is part of the basic concept of what temperature is.* A precise definition of temperature is beyond the

---

* There are a few important situations, such as thermoelectricity and thermodiffusion, in which things other than temperature difference can give rise to a transfer of energy as "heat." See W. C. Reynolds, "Thermodynamics," Chap. 14, McGraw-Hill Book Company, New York, 1968.

scope of this book. Temperature is one of those things that you have to work with for some time without really understanding it before you can come to understand it. Mr. Fahrenheit really didn't understand temperature as it is understood today, but he got on with his work by defining an arbitrary scale of measurement that was good enough for the early days. The development of more scientific concepts and scales of temperature follows the ideas laid out in the last chapter relating to force. Fahrenheit used his body to define 100°F (he had a slight fever) and a chilly day to define 0°F. This corresponds to someone defining force in terms of a rusty spring found in his back yard, with every inch of compression defining the same change in force. A more rational one-point absolute-temperature scale was adopted only as recently as 1954. On the absolute-temperature scale the lowest possible temperature, at which molecular vibration ceases, is zero "degrees Kelvin" (denoted °K). Intervals along the scale are set by defining the "triple point" of water (a peculiar condition where ice, water, and steam exist together) as 273.16°K. The Kelvin scale is the one associated with the metric system (which the entire world will eventually use). Engineers in some countries still make use of another absolute temperature, the *Rankine* scale. A temperature in °R ("degrees Rankine") is $\frac{9}{5}$ of the same temperature in °K. Thus, the triple point of water is at 491.69°R. The Celsius (formerly called "centigrade") and Fahrenheit scales are shifted (nonabsolute) scales corresponding to the Kelvin and Rankine scales, respectively. Fig. 3.4 relates these four

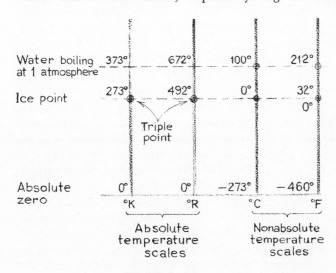

FIG. 3.4 TEMPERATURE SCALES IN COMMON USE.

temperature scales, all of which you are likely to encounter as you read articles about energy.

To summarize, *internal energy* is the energy of atoms and molecules associated with their chaotic microscopic motions. *Heat* is energy transfer from a hot system to a cold one associated with chaotic collisions between their molecules. *Work* is energy transfer between two systems associated with orderly collisions between their molecules (remember the stately parade). *Temperature* is a measure of how hot things are; hotter things can transfer energy as heat to colder things. Now is the time to be sure that the differences between these ideas are clear in your mind.

SOME SIMPLE ENERGY ANALYSES

There is a special class of systems for which the energy analysis is quite simple and straightforward. These are systems that always contain the same matter so that there are no molecules going in or out to carry energy in or out of the system. For a system of fixed matter, the only possible ways to change the energy of the system (macroscopic analysis) are (1) transfer of energy in or out as work, and (2) transfer of energy in or out as heat. To add energy to the system, we can either squeeze it, pull it, twist it, stir it, or shake it (thereby adding energy as work), or we can heat it, say with a flame. To remove energy, we can let it expand, or let it twist, or let it move against a force (thereby removing energy as work), or we can cool it, perhaps with ice. These energy transfers will result in a change in the energy stored within the system; the energy balance connects all these energy terms through proper energy book-keeping, which is usually rather simple for systems in this class.

Often one can choose the system boundaries in a way to carve out such a fixed-matter system from even the most complex device. Consider a large central power station of the type shown in Fig. 1.1. The system shown there is not in this class, because matter flows in and out at several points. But inside such a system there is a closed-circulation loop in which the "working fluid" (usually H_2O) flows around. This circulation loop always contains the same matter, and hence is a fixed-matter system. Figure 3.5 shows a schematic of this fixed-matter system. The water-circulation loop consists of a pump that feeds liquid water into the boiler. In the boiler the liquid water is converted to steam, which then flows to the turbine. The flow leaves the turbine and enters the condenser, in which it is condensed back to a liquid and returned to the pump for recycling. Energy taken from the hot flames (or nuclear fuel) in the boiler is transferred as heat to the steam in the boiler; energy is removed as work output from the steam in the turbine, energy is removed as heat from the

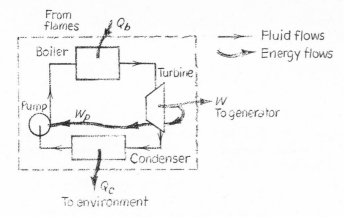

FIG. 3.5 TYPICAL CLOSED-LOOP POWER SYSTEM.

steam in the condenser, and energy is put into the liquid water as work by the pump. In Fig. 3.5 the pump is powered directly within the system by the turbine, and thus W represents the net work that is available for conversion to electrical energy in a generator outside the system.

The energy balance for this system, expressed in terms of the energy flows defined in the figure, and taken over some prescribed time period, is

$$\underset{\substack{\text{energy} \\ \text{in}}}{Q_b} = \underset{\substack{\text{energy} \\ \text{out}}}{W} + Q_c + \underset{\substack{\text{increase in} \\ \text{energy storage}}}{\Delta E_s} \qquad (3.3)$$

The system energy, here denoted by E_s, consists of the kinetic energy of the rapidly rotating turbine disk, the kinetic energy of the high-speed steam shooting out from the turbine nozzle, the potential energy of water, the internal energy of the cold and hot steam, the internal energy of the pipe walls, etc. Once the system has been warmed up and brought into a steady operating condition (termed "steady state") these various terms in E_s become constant; the steam is always just as hot, the turbine disk always goes the same speed, etc.; so there is *no change* in the energy-storage term for steady-state operation. Therefore, $\Delta E_s = 0$, and the energy balance simplifies to

$$\underset{\substack{\text{energy} \\ \text{in}}}{Q_b} = \underset{\substack{\text{energy} \\ \text{out}}}{W} + Q_c \qquad (3.4)$$

Now we can clearly see where the energy put into the boiler is going; part is emerging as useful work and the rest is removed as heat in the condenser. (You may also see why engineers like to "assume steady state"; this gets rid of most of the hard part of the analysis!).

The "cycle-thermal efficiency" of this system, denoted by the Greek symbol η ("eta"), is the ratio of the useful work output to the energy input as heat,

$$\eta = \frac{W}{Q_b} \tag{3.5}$$

Modern plants have cycle efficiencies of the order of 40%; high-technology systems now on the drawing boards run as high as 50%. The overall plant efficiency defined in Chap. 1 will be less than the cycle-thermal efficiency because the boiler will take in more fuel energy than it manages to transfer as heat to the steam and because the generator will convert only about 90% of the turbine work output to electrical energy.

The energy rejected from the system as heat Q_c goes to the local environment, either to a river or to the air. This will warm up the environment, and is the much-discussed "thermal pollution." You know enough now to estimate the amount of energy that is dumped into the environment each year by electrical generation plants (assume an average overall efficiency of about 30%): you'll find it is more than three times the amount of electrical energy generated, and this is why we really do have a thermal-pollution problem. Engineers are now working on ways to utilize the power plant "waste heat" for other purposes, such as residential heating, and this is one of the interesting problems that must receive attention in the years ahead.

Another interesting example of a fixed-matter system is the "heat pump," which is nothing more than a refrigeration system running in a peculiar mode. The system is shown in Fig. 3.6. The compressor circulates the "working fluid," an organic substance called "freon," and also increases the freon temperature during the process of compression. Energy is then transferred as heat from the warmer freon to the relatively cooler

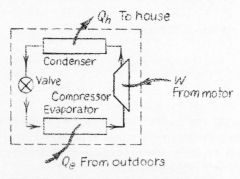

FIG. 3.6 HEAT PUMP FLOW DIAGRAM.

air in the house. This condenses the freon and makes the house warmer. The liquid freon is then passed through a flow restriction (a valve) that causes the pressure to drop. At this lower pressure some of the freon molecules escape from the liquid, forming gas bubbles and taking energy from molecules that remain in the liquid. This dramatically lowers the liquid temperature. Now the freon is cold, so that when it goes through the evaporator it can receive energy as heat from the outside air. This cools off the environment, and so the heat pump in effect is refrigerating the outdoors while it warms the indoors! You probably wonder why go to all this trouble just to heat the house; why not use the electrical energy directly in simple resistance heaters in the house? The answer is given by the energy balance on the system, for which the steady-state condition is (see Fig. 3.6)

$$\underset{\substack{\text{energy}\\\text{in}}}{W \;+\; Q_e} = \underset{\substack{\text{energy}\\\text{out}}}{Q_h} \qquad\qquad (3.6)$$

We see that the term Q_h is *larger* than the compressor energy input W. So we get more energy into the house than we buy from the electric company! And the leverage can be remarkably high; if the temperature outdoors is not too cold, say around 40°F, it is possible that Q_e can be as much as *four times* the compressor work, which means that the energy we get into the house can be *five times* the electrical energy that we use!

There is even an added bonus; with a little bit of extra plumbing we can switch the system around to put the condenser outside and the evaporator inside, and then we can refrigerate the house and heat the outdoors. This gives us a way to cool the house in the summer!

Heat pumps are truly marvels of engineering, and they should see growing applications as our society strives to conserve its energy resources. Figure 3.7 shows a modern household heat pump.

We hope the two previous examples aroused your curiosity about energy-conversion systems. Did you wonder why engineers are so un-clever as to have to throw away 70% of the energy input to the steam-power system or why systems with 80% efficiency are not in operation today? Or did you wonder just how much leverage one could get with the heat pump, why not a factor of 5,000 over the electrical energy input? It turns out that *nature* limits power-system efficiencies to numbers well below 100%. There is an important physical law called the "second law of thermodynamics," and an important concept, "entropy," that apply to such systems. The second law sets a theoretical maximum for the cycle efficiency that cannot be surpassed by even the most clever and sophisti-cated engineer. We'll talk about entropy and the second law very shortly.

FIG. 3.7   GENERAL ELECTRIC HEAT PUMP, AS SET UP FOR LABORATORY DEMONSTRATION AT STANFORD.

HEAT-TRANSFER MECHANISMS

To help you develop a better understanding of heat, let's discuss the process of energy transfer as heat in more detail. There are two important mechanisms by which energy is exchanged as heat between interacting systems. Temperature differences in solids, liquids, and gases produce a flow of energy as heat by processes generally referred to as "conduction." Temperature difference between objects also produces a flow of energy across empty space by the process of "radiation"; radiation can also pass through gases and liquids that are not too dense, such as the earth's atmosphere or the flames in a combustion boiler. In a third process, "convection," moving fluid* transports its internal energy from one region to another, where the energy is removed by conduction or radiation; sometimes convection is called a third mechanism for heat transfer, but it is more precisely regarded as a combination of heat conduction and fluid transport of internal energy.

Some microscopic models of heat-conduction processes will be helpful. We noted earlier that the atoms in solids vibrate frantically at high temperature but are more sedate at lower temperatures. Thus when temperature "gradients" (variations) exist in a solid, the excited molecules bang into the sluggish ones and speed them up, and as a result there is a flow of energy in the direction of lower temperatures. In gases, the higher-temperature regions contain molecules moving rapidly compared with the speeds of molecules in lower-temperature regions. The high-speed molecules fly over to areas of lower temperature, and some low-speed molecules fly the other way, but there is a net flow of energy from the higher-temperature regions to regions of lower temperature. The process of conduction in a liquid is somewhere between those in a solid and those in a gas.

In order to give you a better feel for energy technology, it will be helpful to review the basic mathematical formulations used to describe energy transfer as heat. This won't make you a heat-transfer engineer, but it may help you understand one! Let's first talk about conduction. The rate at which energy is transferred as heat by conduction (e.g., the calories per second) is proportional to the temperature gradient; if the temperature varies quite sharply over a short distance, there will be a high rate of energy transfer. The conduction heat-transfer rate is also proportional to the cross-sectional area that carries the energy flow—the larger the conducting area, the more the energy that flows through it. This behavior is described by the "heat-conduction-rate equation," sometimes called "Fourier's law,"

$$\dot{Q} = \frac{kA(T_1 - T_2)}{d} \tag{3.7}$$

We have written it here for the special case of conduction between two surfaces separated by a slab of material (Fig. 3.8). T_1 and T_2 are the temperatures of the two surfaces, A is the cross-sectional area (area perpendicular to the direction of energy transfer), d is the distance between the two surfaces, k is a property of the conducting material called the "thermal conductivity," and $\dot{Q}$ is the rate of energy transfer as heat between the surfaces (e.g., Btu/hr or cal/sec). (Throughout the book we

* "Fluid" means gas or liquid.

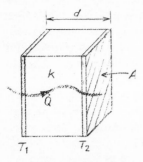

FIG. 3.8 TERMS IN THE
CONDUCTION-RATE EQUATION.

use overdots to represent rates of flow.) Note that, for a given material
and temperatures, the rate of energy transfer as heat will be greatest
when the thickness d is smallest; this is why insulation is made thick and
frying pans are made thin. Since $\dot{Q}$ has the dimensions of power, the
thermal conductivity k will have a messy set of dimensions; you can
figure these out from Eq. (3.7) and check your result with Table 3.1.
The table shows that copper and aluminium are among the best conductors of heat; for a given thickness and temperature difference, more
energy will flow as heat through copper than through any other known
substance. Engineers take advantage of this whenever the conduction
process must be used to remove energy as heat quickly from a source;
for example, transistors are mounted in copper to provide for rapid
transfer of the energy as heat away from semiconductor material that
could be damaged by becoming too hot.

Suppose we are interested in determining the rate of heat transfer
through the wall of a house. Note that the air spaces in the walls will
have a low thermal conductivity, and that the wood is a better conductor
(the nails are even better but don't provide much conduction area!). Wood
accounts for perhaps 10% of the conduction area in a typical frame wall,
so let's estimate that the average thermal conductivity of the wall will be

Table 3.1

Thermal Conductivities of Various
Materials

Material	Thermal conductivity k, Btu/hr-ft-°F
Room air	0.014
Wood	0.06–0.08
Tap water	0.35
Brick	0.4
Steel	6–20
Aluminum	120
Copper	220

about 0.02 Btu/hr-ft-°F. This might be quite close if the walls are stuffed with some light fiber-glass insulation to prevent convective motions of air within the walls from contributing significantly to the energy transfer through the wall. Now, walls are about $\frac{1}{4}$ ft thick, and we'll look at an area of 1 ft^2. Then, if the inside temperature is 70°F and the outside is 30°F, which correspond to 530°R and 490°R, respectively, Eq. (3.7) gives

$$\dot{Q} = 0.02 \text{ Btu/hr-ft-°R} \times \frac{1 \text{ ft}^2}{0.25 \text{ ft}} \times (530 - 490)°R = 3.2 \text{ Btu/hr}$$

(Note that we could have used the temperatures in °F to compute the temperature difference.) Thus, for each square foot of wall area there would be a heat-transfer rate to the outside of 3.2 Btu/hr. At windows the rate will be much higher and in fact will be determined by the convective processes inside and outside rather than by conduction in the glass. A good house designer will make calculations like these to estimate the total rate of energy transfer as heat from the house to the outdoors on the coldest day, and then be sure that the house heating system can supply at least this much energy to the air inside the house to keep the house at a steady, comfortable temperature.

Convective heat transfer is described by an equation similar to Eq. (3.7),

$$\dot{Q} = hA(T_2 - T_1) \tag{3.8}$$

Here $T_2 - T_1$ is the appropriate temperature difference, h is the "convection heat-transfer coefficient," and A is the area of the surface exposed to the convection process (Fig. 3.9). The value of h depends on the nature of the convecting flow. Faster flows tend to be quite turbulent (see Fig. 3.9), and this markedly enhances the ability of the flow to pick up energy from the solid surfaces and convect it away. Slower flows can be very smooth, or "laminar," and without turbulence the convection heat-transfer coefficient will be much lower. Convective flows can either be "forced" by a blower that makes the flow move, or arise naturally ("natural" or "free" convection) as a result of the buoyancy of warm fluid surrounded by cooler fluid.

Values of h for different conditions are sometimes derived using complicated mathematics and sometimes by experiments. A large body of literature now exists on this subject, created almost entirely by basic research in heat transfer carried out by engineers in universities, industry, and government research centers. Pioneering work on heat transfer and related fluid-flow problems continues to be a very important aspect of graduate programs in mechanical, aeronautical, and chemical engineering departments at our major universities.

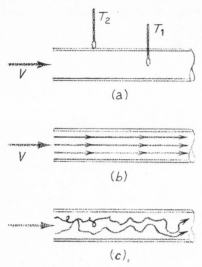

FIG. 3.9 (*a*) FOR CONVECTION
HEAT TRANSFER TO A FLUID
FLOWING IN A PIPE, T_1 IS THE
AVERAGE FLUID TEMPERATURE, T_2
THE WALL TEMPERATURE, AND A
THE SURFACE AREA INSIDE THE
PIPE. (*b*) AT LOW SPEEDS THE
FLUID MOVES SMOOTHLY THROUGH
THE PIPE IN "LAMINAR" FLOW.
(*c*) AT HIGHER SPEEDS THE FLOW
MOVES IRREGULARLY IN
"TURBULENT" FLOW.

"Heat exchangers" are important components of many energy systems.
These are devices used to transfer energy as heat between two fluids.
There are many schemes, all involving surfaces (such as pipes) that
separate the two fluids. Energy is transferred by convection to the pipe
from the hotter fluid, then through the pipe by conduction, and then to the
colder fluid by convection. Often "fins" are added to the outside or inside
of the pipes to provide more surface area for the convection process; an
automobile radiator is an example of a "finned-tube heat exchanger."
The development of heat exchangers that are very small but have a high
energy-transfer capability is an example of "high energy-technology."
Such "compact heat exchangers" incorporate a lot of heat-transfer area
and are cleverly designed to produce high values of the convection heat-
transfer coefficient. Very novel, compact exchangers, in which the surfaces
are themselves moved between the hot and cold fluids, using ceramic
materials that survive at very high temperatures, are now used in vehicular
gas-turbine engines. Figure 3.10 shows some of the heat-exchanger surfaces
used in modern heat exchangers.

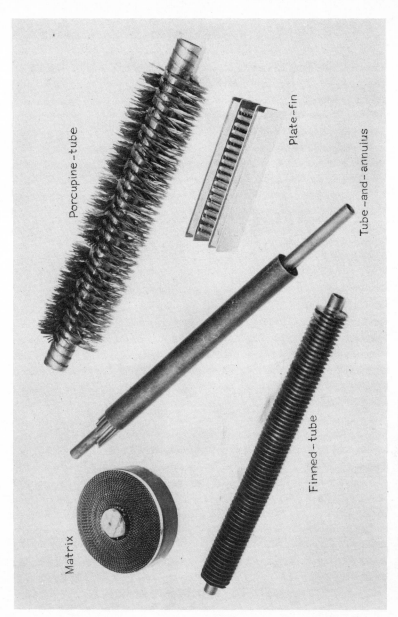

Porcupine-tube

Plate-fin

Tube-and-annulus

Matrix

Finned-tube

FIG. 3.10 HEAT-EXCHANGER SURFACES.

Nuclear power reactors consist of fuel plates or rod bundles in which the nuclear reaction releases energy. This energy is conducted to the outside surface of the fuel elements, where it is transferred by convection to a fluid flowing past the fuel element (see Fig. 3.11). Various tricks such as flow swirlers and turbulence promotors have been used to increase the convection heat-transfer coefficient and thereby obtain a higher power output for given temperature difference.

The basic physics and describing equations of radiation heat transfer are quite different from those for conduction or convection. Every surface emits radiation energy at a rate that is proportional to the fourth power of its absolute temperature. The rate of heat radiation from a surface is described by

$$\dot{Q} = \varepsilon\sigma A T^4 \tag{3.9}$$

where A is the area of the radiating surface, T is its *absolute* temperature (°K or °R), σ is the "Stefan-Boltzmann constant" $= 1.72 \times 10^{-9}$ Btu/hr-ft^2-°R^4, and ε is a dimensionless property of the surface called the "emissivity." Surfaces with $\varepsilon = 1$ are called "black bodies"; they radiate the maximum amount of power for a given temperature. Oxidized metal surfaces have emissivities of the order of 0.6 to 0.9, while shiny surfaces have smaller emissivities, perhaps as low as 0.01. Heat can be removed from space vehicles only by radiation, and surfaces designed to do this will be chosen to have a high emissitivity. In Thermos-type vacuum bottles one is instead interested in minimizing the amount of radiated energy, which is why shiny surfaces are used.

Radiation can be thought of as the emission of photons (or light) at a variety of wavelengths; but not all radiation is visible. For example, the technique of infrared photography, which has been used to locate warm material in situations ranging from jungle warfare to power station thermal pollution, is based on nonvisible radiation. Heat radiation is just a form of electromagnetic radiation; other forms include television and radio transmissions. However, the radiation from a radio transmitter is *organized* (that is how it contains information), while heat radiation is disorganized; it contains no "signal," just "noise." This is because the photons are emitted randomly in random directions and with a broad band of energies.

When radiation falls on a surface, some of it is reflected, some is absorbed, and some might be transmitted through the surface. Solid surfaces are quite opaque to radiation, and hence either reflect or absorb the incident energy. The radiation energy incident on the earth from the sun is partially absorbed by the atmosphere and partially reflected, and

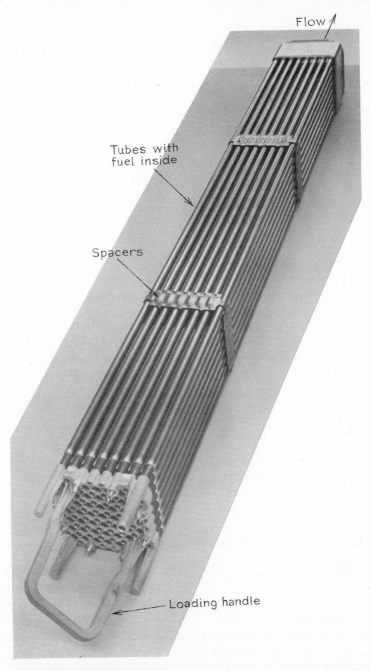

Flow ↗

Tubes with
fuel inside

Spacers

Loading handle

FIG. 3.11 NUCLEAR REACTOR FUEL ELEMENT. COURTESY OF THE GENERAL
ELECTRIC COMPANY.

quite a lot gets through to the earth's surface. The solar radiation incident upon the atmosphere is about 1.4 kw/square meter of surface area perpendicular to the direction of radiation; this gives you a figure which, together with the diameter of the earth, you can use to calculate the amount of solar power incident upon the earth as about 1.7×10^{17} watts.

Two surfaces facing each other radiate energy toward one another and will reflect and absorb each other's energy. The net energy exchange by radiation between the two surfaces depends in a complicated way upon the surface geometries, upon the absorption, reflection, and emission characteristics of each, and upon the fourth powers of their two absolute temperatures. The resulting heat-transfer-rate equation is of the form

$$\dot{Q} = \sigma F A (T_1{}^4 - T_2{}^4) \tag{3.10}$$

Here A is the area of either surface, F is a factor that includes the complexities just described, and T_1 and T_2 are the absolute temperatures of the two surfaces. If you recall your algebra, you can factor the temperature term and convert Eq. (3.10) to

$$\dot{Q} = h_r A (T_1 - T_2) \tag{3.11}$$

where the "radiation heat-transfer coefficient" h_r is a complicated expression involving σ, F, T_1, and T_2. When T_1 and T_2 are close to "room temperature," h_r will be of the order of 1 Btu/hr-ft²-°R for typical radiation situations. You can use this number to estimate the radiant exchange between your body and the environment. Your surface area is probably of the order of 15 ft² (apologies for any insults!). Your skin temperature is about 556°R (96°F); suppose that the surroundings are about 500°R (40°F), so that

$$\dot{Q} = 1 \text{ Btu/hr-ft}^2\text{-}°R \times 15 \text{ ft}^2 \times (556\text{-}500)°R$$

$$= 840 \text{ Btu/hr}$$

This of course presumes you are standing nude in the shade. Since the h for natural convention is also about 1 Btu/hr-ft²-°R, about the same amount would be removed by natural convection, giving you a total energy output as heat of about 1,600 Btu/hr. Your body wants to reject only about 500 Btu/hr, and so you wear clothes to reduce the total heat-transfer coefficient between your skin and the environment. You can calculate for yourself the outdoor temperature for which you would be comfortable in the nude in the shade.*

To summarize, the rate of energy transfer as heat between two systems is determined by the mechanisms of heat transfer, by the geometry of the system and the properties of the media involved, and by the magnitude of the temperature difference. The larger the temperature difference, the greater the heat-transfer rate.

AVAILABILITY, IRREVERSIBILITY, AND ENTROPY

We are now ready to discuss the very basic ideas behind the answers to questions about the "best possible" energy systems. Consider a system containing a lot of energy at a high temperature. Suppose we inadvertently or deliberately allow this system to interact with a system at room temperature. Energy transfer as heat will take place by one or more of the processes just described, and will result in the transfer of energy from the warmer system to the cooler system. Something has been lost in the process, but what? We have not lost any energy; the energy has just moved from one place to another. But the hot energy could have been used to make steam in a power plant, and the steam could have been used to make useful work. We could have generated some badly needed electrical energy, but we didn't. Now that the energy resides in a cold region we can't make steam and therefore we can't make useful work. Not only can't we use the cold energy to generate electricity, we can't even use it to melt lead; by moving the energy from the higher-temperature system to the lower-temperature one we "lost" some ability to use the energy constructively; we lost some possibility to do work! In scientific terms, the "availability" of the energy has been reduced.

Suppose that we decide we must return the energy to the hot system. To do so will require a "heat pump" of the type described earlier in the chapter. The heat pump will in turn require an input of work. So, we will have to expend badly needed energy to repair the original loss in the availability of the energy in the hot system. However, in so doing we must use up available energy (work) to run the heat pump; i.e., we *reduce* the available energy in the environment. Thus, overall we don't gain any available energy by trying to restore the original conditions in the hot system; we are stuck with a permanent loss of available energy.

Consider another system, this time a rapidly rotating flywheel immersed in a tank of air. Initially there is a lot of energy in the speeding

*This calculation neglects evaporative cooling, which is important to human comfort.

flywheel; this energy could be used to generate electricity, to lift loads, or even to move a car. The friction with the air will gradually slow down the flywheel, and once it has come to rest, we will have lost all possibility of doing useful work with the flywheel kinetic energy. Again there has been no loss of energy; it has simply moved from the form of flywheel kinetic energy to internal energy in the air and steel. But there has been a loss of *available* energy, just as before.*

There are many other processes that lead to loss of available energy, and you can think up examples of each. They include mixing of two fluids, spontaneous chemical reactions, all kinds of friction, electric current flow through resistors, and more (how about death?). These processes are called "irreversible" because they result in a permanent and irretrievable loss of available energy. Engineers call the amount of available energy loss the "irreversibility" of the process, and good engineers are usually on guard against unnecessary irreversibility. Some irreversibility is almost always necessary in real systems. For example, in order to make a lot of energy to flow as heat from flames to steam through a heat-transfer area of reasonable size, a large temperature difference is required [see Eq. (3.8)]; the dropping of energy from flame temperature to steam temperature is an irreversibility that must be accepted if one wants to have a boiler of finite size and cost. So the engineer is always faced with a compromise and must try to balance the disadvantages of irreversibility with other factors.

Some processes have only a very small degree of irreversibility, and these are sometimes modeled as being "reversible." Figure 3.12 shows one such example. The spring-mass system can be set into oscillatory motion, and if the block is well lubricated, the friction can be reduced to the point where many cycles of oscillation will occur without noticeable reduction in the enthusiasm of the oscillatory motion. In the theoretical limit of zero friction the oscillatory motion will go on forever. There is no loss of available energy at any time during the oscillation, which is why the theoretical limiting process is reversible. Once each cycle the system returns exactly to the point where it was before, and thus whatever happened to the system during the cycle has been reversed. During the oscillation energy passes back and forth between the kinetic energy of the mass and the strain (internal) energy of the spring, but never (ideally) is there any

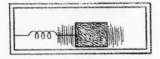

FIG. 3.12 THE OSCILLATION WILL BECOME REVERSIBLE AS FRICTION VANISHES.

* Some recent writers use the term "exergy" for available energy.

loss of available energy. The concept of the reversible process as a theoretical limit for some (but not all) real processes is very useful in the science of thermodynamics.

In the flywheel discussed above, the initial state of affairs saw the molecules in the flywheel revolving around the axis in a stately parade; this is a very *organized* situation, and it is this molecular organization that makes the flywheel energy fully available. In the slow-down process the organized energy is gradually randomized until it all appears in the randomly oriented motions of individual steel and air molecules. The energy is no longer available to do useful work because the organization that we could use has disappeared. *Randomization at the microscopic level always characterizes irreversibility.*

When the flywheel is rotating, we *know* that the molecules are moving around the axis in a particular direction. This knowledge allows us to capture their kinetic energy. When the wheel comes to rest, the molecules in the steel and air have taken on the original rotational energy of the wheel, but the molecules now move in all different directions at once. The lack of knowledge of what each molecule is doing at each instant prevents us from capturing their energy. *Loss of information about the state of things at the microscopic level always characterizes irreversibility.*

These ideas about irreversibility, loss of available energy, and microscopic randomness and uncertainty about the microscopic state are expressed quantitatively in terms of a property of matter called "entropy" (en-tro-pee). Entropy, like energy, is a conceptual quantity that one must come to understand by talking about it and working with it; any equations that are proposed to define it have to be fully consistent with the ideas about it. There are four basic aspects to the concept of entropy:

1 All systems have entropy, which measures the disorganization at the microscopic level.
2 The entropy of the whole is the sum of the entropy of the parts.
3 The entropy of a perfectly ordered system will be zero.
4 Entropy can be produced but can never be destroyed.

The first expresses the ideas discussed above. The second tells us that the amount of entropy in a complex system must be the sum of the entropies of its parts. The third is important in considering matter at absolute zero temperature; pure substances at absolute zero are fully ordered, and hence must have zero entropy. The fourth is the famous (or infamous?) "second law of thermodynamics." Note that, while *energy* is neither produced nor destroyed, *entropy* can be produced, but it can *never* be destroyed. The second law of thermodynamics provides the mathematical structure within which engineers and scientists quantitatively deal with questions about maximum efficiencies of power plants, minimum work requirements for various processes, and the chemical composition of the products of a chemical reaction.

Various expressions for the entropy of matter appear in books and papers on the subject. One famous one is

$$S = k \ln \Omega \tag{3.12}$$

where k is a basic physical constant called "Boltzmann's constant," Ω is the number of possible microscopic configurations of the system, and $\ln \Omega$ denotes the "natural logarithm" of Ω. [Equation (3.12) is the sole inscription on Boltzmann's gravestone!] For a fully ordered system $\Omega = 1$, and so $S = 0$ as required above. Systems with many possible microscopic states will jump from one state to another, and the more possible states there are the bigger is S. And, if there are Ω_A microscopic states possible in system A, and Ω_B in system B, then in the combined system $A + B$ there must be $\Omega_A \times \Omega_B$ microscopic states possible. If you remember how logarithms work, you will easily see that Eq. (3.12) gives $S_{A+B} = S_A + S_B$ as required above. Clearly Eq. (3.12) meets three of the four requirements outlined above.

Let's use an analogy to see how Eq. (3.12) relates to the "second law of thermodynamics," as described above. Imagine that you have 100 red jumping beans and 100 white; place the red ones on one side of a tray and the white ones on another. There is only one way to do this (all red on one side and all white on the other); so $\Omega = 1$ for this configuration and hence $S = 0$ (recall $\ln 1 = 0$). Now let them jump. After a while the red and white ones will be spread all over the tray, and the tray will look "pink" rather than half-red and half-white. There are many ways you could arrange the beans so that the tray looks pink. So Ω must be very large, and hence S has gone from zero to some larger number; the entropy has indeed increased!

Equation (3.12) does fit all the requirements on entropy, but only for systems in which each microscopic state is equally probable. If the various microscopic states occur with different probabilities, then scientists make use of the expression

$$S = -k \sum_i p_i \ln p_i \tag{3.13}$$

Here p_i is the "probability" (a concept we shall not discuss) of the ith microscopic state and k is again Boltzmann's constant. In mathematics the Greek letter $\sum$ ("sigma") means "sum what follows over all values;" so Eq. (3.13) means

$$S = -k(p_1 \ln p_1 + p_2 \ln p_2 + \cdots + p_\Omega \ln p_\Omega)$$

If each of the Ω states is equally likely, $p_i = 1/\Omega$ and Eq. (3.13) reduces to Eq. (3.12). Since probabilities are less than 1, and logarithms of numbers between 0 and 1 are negative, S is positive, in spite of the minus sign in Eq. (3.13).

You may have run across Eq. (3.13) in some other areas. It has been used to define "entropy" in many other situations, most notably in the mathematical theory of information and communication. As social scientists become more quantitative, they may begin to think in terms of the "entropy of a given society" and perhaps even to measure this entropy with a definition like Eq. (3.13). This general concept of entropy is also a key factor in modern theories of decision making.

Practical engineering analysis requires knowledge of the entropy of all sorts of materials at all sorts of pressures and temperatures. While a good physicist can use the microscopic definitions Eqs. (3.12) or (3.13) to compute S for some simple cases, most of the entropy data have been generated by an entirely different macroscopic approach. This was the original approach to entropy introduced by the thermodynamicists of the nineteenth century, which is still used in most textbooks on thermodynamics. They state that, for a *reversible* process, the change in the entropy of a fixed-matter system is equal to the amount of energy that the system *receives* as heat divided by its *absolute* temperature (°R or °K). In terms of the symbols in Fig. 3.13,

$$\Delta S = \frac{Q}{T} \qquad \text{reversible process} \tag{3.14}$$

It is probably not at all obvious that Eq. (3.14) has any relationship to the ideas about microscopic disorder discussed above, but the connection can indeed be made. Q represents energy input as heat, which we can think of as *disorganized energy transfer* (as opposed to work, an organized

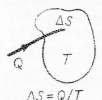

FIG. 3.13   RELATING ENTROPY CHANGE TO ENERGY TRANSFER AS HEAT.

transfer of energy). Doesn't it seem that disorder should flow into a system if we transfer energy into it in a disorganized manner? If we let some energy leak out (negative Q) through a disordered transfer process, doesn't it seem reasonable that disorder should flow out? Now, according to Eq. (3.14) the colder the system the greater the change in entropy for a given energy transfer Q. We can explain this microscopically by arguing that in very cold solids there is not much molecular excitation and therefore adding a little bit of energy will stir things up quite a bit, adding lots of molecular disorder. At high temperatures things are already rather disordered, so that a few extra Btu do not add much entropy. The connections between Eqs. (3.12), (3.13), and (3.14) can be made much more rigorously; in fact, the Boltzmann constant k in Eqs. (3.12) and (3.13) is put there for the specific purpose of making these equations produce the same results as Eq. (3.14) in theoretical treatments of the entropy of ideal gases.

We hope you have taken the time to figure out the dimensions of entropy from Eq. (3.14); energy/temperature. Common units for entropy are

Cal/°K, erg/°K, joule/°K, Btu/°R, ft-lbf/°R

The big virtue of Eq. (3.14) is that it provides physical chemists with a means for measuring changes in entropy in the laboratory; one has only to measure the energy transfer as heat Q and the temperature T. The stickler is that the process must be reversible, for if it is not, the entropy change will *not* be equal to Q/T. Since all processes are somewhat irreversible, one has to work very hard to reduce the irreversibility to the point where Eq. (3.14) provides an adequate evaluation of ΔS. Fortunately this is possible, and the entropy of practically every material ever invented has been measured over wide ranges of conditions by diligent laboratory thermodynamicists and tabulated for use in engineering analysis.

Armed with the second law and values for the entropy of all substances, the engineer can figure out what can and cannot be done with a system of interest. For example, he might want to compress air; the second law (together with the energy balance) gives him the tools to calculate the minimum work that is required for the process, and he can be sure that no earthly inventor will *ever* make a machine better than this theoretical ideal! The power of thermodynamic theory is very great, for one can find limits on what can be done without any reference to the details of particular schemes!

The "best" or "ideal" process is the one that is "reversible," i.e., it does not produce any entropy, i.e., it does not destroy availability, i.e., it has zero irreversibilities. Frictionless motion is a reversible process; reversible processes are ideals that might be approached very closely but are never met.

In carrying out available-energy analysis, engineers have developed some other helpful ideas. Equation (3.14) can be interpreted as saying that, whenever energy is transferred as heat across the boundaries of a system, there is an associated "entropy flow." If the energy is put into the system as heat, there is an entropy inflow; if the energy is put out as heat, there is an entropy outflow. The amount of "entropy flow with heat" across the system boundary is

$$S_{in} = \frac{Q_{in}}{T_{boundary}} \qquad\qquad (3.15a)$$

or

$$S_{out} = \frac{Q_{out}}{T_{boundary}} \qquad\qquad (3.15b)$$

depending on the direction of energy transfer as heat. Equations (3.15) describe the entropy flow for any process, reversible or irreversible. (Don't forget that T represents the *absolute* temperature, °R or °K.) For a reversible process there is no entropy production within the system, and so the system entropy change exactly balances the entropy flow, as prescribed by Eq. (3.14).

Equations (3.15) allows us to evaluate the amounts of "entropy flows with heat." What about "entropy flows with work"? Work is, by conception, fully organized energy transfer, and hence one might correctly guess that there is *no* entropy flow across a system boundary associated with energy flow as work. Indeed, truly precise discrimination between heat and work rests on the concept that entropy flows only with heat, never with work. So, to fully comprehend the concepts of heat and work, one really must simultaneously comprehend the concept of entropy! It takes years for persons working in thermodynamics to reach this deep level of comprehension; so you should not be troubled if at this point you have some uneasiness with these ideas.

These ideas have one particularly important application with which you should become familiar—the "Carnot efficiency." This is the efficiency of a "Carnot engine," which provides an upper limit for the performance of all real heat engines (such as that in Fig. 3.5). Figure 3.14 shows an engine taking in energy as heat Q_1 from a region at temperature T_1, converting some of this energy to work W, and rejecting some

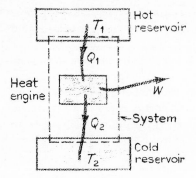

FIG. 3.14   ANALYZING THE "BEST POSSIBLE" HEAT ENGINE.

energy Q_2 to a region at temperature T_2. Let's analyze the system enclosed by dotted lines and make an entropy accounting. Entropy flows into the system from region 1 in the amount

$$S_{in} = \frac{Q_1}{T_1} \qquad\qquad (3.16a)$$

and out to region 2 in the amount

$$S_{out} = \frac{Q_2}{T_2} \qquad\qquad (3.16b)$$

There is no entropy flow associated with the work W. So the change in the entropy of the system is given by bookkeeping as

$$\Delta S_{system} = S_{in} - S_{out} + S_{produced} \qquad\qquad (3.17)$$

where $S_{produced}$ represents the entropy *produced* by irreversible events inside the engine. For steady-state operation of the engine, the amount of entropy stored within the system does not change as time passes; so

$$\Delta S_{system} = 0 \qquad\qquad (3.18)$$

Combining the equations above, we can rewrite Eq. (3.17) as

$$S_{produced} = \frac{Q_2}{T_2} - \frac{Q_1}{T_1} \qquad\qquad (3.19)$$

So far, all we have done is entropy bookkeeping, Now we invoke the second law of thermodynamics, which requires that the entropy production can't possibly be negative. So

$$S_{\text{produced}} = \frac{Q_2}{T_2} - \frac{Q_1}{T_1} \geq 0 \qquad (3.20)$$

or

$$\frac{Q_2}{Q_1} \geq \frac{T_2}{T_1} \qquad (3.21)$$

So, for a given heat input Q_1, the heat rejection Q_2 must at least be as large as $Q_1(T_2/T_1)$. We can relate Q_2 to Q_1 and the work W through an energy balance on the system,

$$W = Q_1 - Q_2 \qquad (3.22)$$

So the engine thermal efficiency is

$$\eta = \frac{\dot{W}}{Q_1} = \frac{Q_1 - Q_2}{Q_1} = 1 - \frac{Q_2}{Q_1} \leq 1 - \frac{T_2}{T_1} \qquad (3.23)$$

We see that the second law gives us an expression for the *maximum* thermal efficiency of heat engines, without any consideration of the makeup of the engine! This maximum efficiency is called the "Carnot efficiency," after the original inventor of entropy (who interestingly enough did not really understand energy!):

$$\eta_{\text{Carnot}} = 1 - \frac{T_2}{T_1} \qquad (3.24)$$

Equation (3.24) suggests that, in order to obtain high efficiencies, one wants to have low T_2 and high T_1; T_2 will be determined by the environment temperature, which at 60°F corresponds to $T_2 = 460 + 60 = 520°R$ (*don't forget that T here is absolute temperature!*). T_1 will be limited by materials, and temperatures of 2000°R are not unrealistic. For these two temperatures

$$\eta_{\text{Carnot}} = 1 - \frac{520}{2,000} = 0.74$$

So 74% would be the maximum thermal efficiency of power plants operating between these temperatures. Typically, a practical plant can develop half this ideal efficiency; so it might have an actual efficiency of 37%.

In order to come closer to the Carnot efficiency, engineers must somehow reduce irreversibilities all through the system. Research in such areas as the fluid mechanics, combustion, heat transfer, and lubrication has led to slow but continual improvements in operating efficiencies. Efficiencies will also improve as high-temperature metallurgy permits higher and higher temperatures to be employed in thermal power systems.

It would be well to outline what we expect from you at this point. We hope that you understand that entropy is a measure of microscopic disorder, of big old man's uncertainty about what all those little tiny molecules are doing. You should know that the second law of thermodynamics is a general law of nature that describes goings-on at the microscopic level, where things naturally tend to get more and more disorganized. Man's inability to control the actions of individual molecules makes it tough for him to utilize their energy fully. The natural irreversibilities of all real processes cause a loss of the availability of the energy. You should be aware that a well-trained engineer can, by careful energy and entropy bookkeeping, calculate the very best that can be done in any process, and thereby set his sights and hopes accordingly. If you yourself want to be able to work confidently with these ideas and procedures, you'll have to take a comprehensive course in applied thermodynamics.

PROBLEMS

To develop skill with numbers

3.1 A system receives 40 Btu of energy as heat and 75 Btu of energy as work. The initial internal energy was 139 Btu. Calculate the final internal energy.

3.2 A system receives 90 Btu of energy as heat while rejecting 140 Btu of energy as work. The initial internal energy was 720 Btu; calculate the final internal energy.

3.3 A flywheel spins in a pot of oil. The initial kinetic energy of the flywheel was 10,000 ft-lbf, and the initial internal energy of the wheel and oil was 50 Btu. Calculate the final internal energy of the wheel and oil after equilibrium has been obtained.

3.4 For the heat pump of Fig. 3.6, Q_e is 100,000 Btu each hour, and the motor draws 10 hp. Compute the energy input to the house each hour, in Btu and in kwhr.

3.5 A power plant produces 350 Mw of electrical energy from nuclear fuel with an overall plant efficiency of 35%. Calculate the energy-transfer rate to the river-water coolant in Btu/hr. How many gallons of water would this power elevate 10°F in an hour?

3.6 A substance melts at 500°R with an energy input as heat of 40 Btu/lbm. The process may be treated as reversible (you can freeze the liquid by removing the same amount of energy as heat). Calculate the entropy change for the melting process.

3.7 Calculate the maximum possible cycle-thermal efficiency of a heat engine operating between cold, nonpolluting flames at 300°F and an environment at 60°F. What would you think of an inventor who claimed to have a magnetosterofluidozincal engine that could convert 45% of the flame energy into useful work?

3.8 Calculate the rate of heat conduction through a copper bar 1 in square and 10 ft long if the end temperatures are 250°F and 100°F.

3.9 In a particular gas-cooled nuclear reactor the convection heat-transfer coefficient is 250 Btu/hr-ft^2-°F, and the temperature difference between the fluid and solid is (on average) 400°F. Calculate the amount of surface area necessary to allow a total heat-transfer rate of 3 million Btu/hr.

3.10 A space vehicle must radiate 10,000 Btu/hr to free space. Assuming that the radiator surface is at 400°R, and that the surface emissivity is 0.8, calculate the radiator area required, assuming the radiator is shaded from the sun.

Some more interesting analyses

3.11 Practical fossil-fuel power plants operate at about half the Carnot efficiency. Suppose that the air-pollution experts tell us that the combustion products from a 500°F burner are only one-third as dirty as those from a 1500°F burner per Btu of fuel energy. For a given electrical-power output, which burner temperature will result in the lower total production of pollutants?

3.12 It has been proposed that power plants be built to take energy as heat from warm water near the surface of the ocean (27°C) and reject energy as heat to the cold water deep below (19°C). What maximum efficiency could be obtained with such systems? How would you react to a proposal to spend 40 billion dollars in development of a solar sea-energy program?

3.13 It has been proposed that natural-gas water heaters be banned for air-pollution reasons and, instead, that water in houses be heated with electrical energy generated at a distant power plant. The power station burns natural gas and produces electrical energy with 40% efficiency; the transmission line loses 10% of the electrical energy as heat. All that remains gets into the water. In contrast, the gas water heater gets only 60% of the fuel energy into the water. Which scheme uses more natural gas? Discuss the other factors that would be important in deciding whether or not to support a proposed law banning gas water heaters.

3.14 A new city is being planned for an undeveloped area. A 300-Mw nuclear power plant will be constructed near the city, and the engineers planned to use cooling towers to remove the "waste heat" from the power station. An environmentalist group has attacked this plan, and instead proposes to use the waste heat for space heating in the homes and offices in the city. The demand for electrical energy will be fairly uniform over the year at about 4 million kwhr/day. The space-heating demands will vary from about 25×10^9 Btu/day in the summer to about 300×10^9 Btu/day in the winter. An underground hot-water tank could be used to store excess "waste heat" to smooth out the space-heating demands for periods of a few days at most. Hot water generated at the power plant would be piped to the city for space heating; natural-gas water heaters at the power plant could provide additional heating of the water if the power-plant heat rejection is insufficient. The power-plant efficiency will be about 35% if cooling towers are used, but only about 30% if the waste heat has to be rejected to the recirculating space-heating water system.
Examine the energetics of this system, and critique the proposal for utilizing the waste heat for space heating. What additional arrangements will be necessary to provide for mismatch of the electrical and heating demands? Can you find a compromise scheme that might satisfy the values of both groups?

3.15 A proposed solar power system will collect energy radiated from the sun, use this energy to generate steam, and reject heat to the atmosphere through cooling towers. Using parabolic mirrors to focus the radiant energy on pipes, temperatures of 1,000°F can be achieved. For locations in the southwestern United States the average energy that can be collected is about 200 Btu/hr/square foot of earth surface area. The solar energy will be used to generate steam for a conventional power plant having an efficiency of 35%.
A group opposed to nuclear power has been pressing the President to set a national goal of developing solar energy along these lines in a massive way over the next decade. As a staff analyst in the Office of Technology Assessment, you have been asked to make a quick preliminary analysis and provide some basic input. Suppose that the new electrical-power capability the country will need to install in the next 10 years is developed using the solar-energy scheme in question. How much land area will be required? If each plant has an installed capacity of 1,000 Mw, how many plants will be required, and what land area would be associated with each? How might the problem of nighttime power production be handled? (See Prob. 2.14.) What other technical and nontechnical factors should be considered in a more comprehensive analysis of the problem? (What problems might birds cause?)

4

FLOW SYSTEM ENERGETICS

In which we glimpse
applied thermodynamics
doing its thing

THE WHYS AND WHATS OF WORKING FLUIDS

Many important systems use a "working fluid" to carry out various energy-conversion functions. The steam power plant of Fig. 1.1 and the heat pump of Fig. 3.7 are examples of such systems. A first step in the design of such a system is preliminary thermodynamic analysis in which the engineer uses a combination of scientific principles, data about the properties of the working fluid, and estimates of various component performance parameters to predict the behavior of the system. This process is called "thermodynamic analysis." It tells the engineer the temperatures and pressures that will occur in the system and the amount of working fluid that must be circulated per hour to do the job at hand, and it gives one the data needed for first-cut estimation of the size (and hence cost) of the various pumps, heat exchangers, compressors, turbines, piping, etc., that make up the system. A basic understanding of this process will be helpful to you in reviewing new or proposed energy systems and will facilitate your reading of the literature on energy systems; this chapter is included with this objective in mind.*

Why use a fluid at all? This is a good question, and there is a great deal of interest today in "direct-energy-conversion" systems that can convert energy from one form to another without use of an intermediate working fluid and the pumps, heat exchangers, compressors, turbines, etc., that fluid systems employ. For example, solar cells used on space vehicles are direct-energy-conversion systems; their light weight makes them well suited for space application, but their low power output and poor efficiency makes them losers on earth in comparison with conventional energy-conversion systems. Research in materials science and technology may someday bring direct-energy-conversion systems up to the performance levels of fluid systems, but fluid systems will always remain a large and important factor in energy technology. So, to know about energy technology, we must know about the energetics of fluid systems.

To know about fluid systems, we must know something about fluids. You are probably familiar with a few, such as air, water, steam, milk, kerosene, and beer, but these are only some of the important fluids in energy technology. Man-made organic fluids called freons are very important in refrigeration systems. Liquid metals such as sodium, potassium, a mixture of the two called NAK, and cesium are very important in high-temperature energy technology. Each fluid has its own peculiar characteristics that make it well-suited for a particular job, and each has its problems that the engineer must work around. For example, some freons are quite poisonous, and the liquid metals are highly corrosive at high temperatures. Even air and water have their problems.

In selecting the fluid for a particular application, attention is often paid to the boiling properties of the liquid. You probably know that water boils at 212°F, but did you know that this is true only if the pressure of air or steam around the water is one atmosphere (14.7 psia)? If we want to boil water at a lower temperature, say 80°F, the pressure would have to be reduced to about 0.25 psia. At this low pressure a pound of steam takes up almost fifty times as much space as 212°F steam, which means that large pipes will be required to carry the steam. And, if a joint in the

---

* You can probably understand most of the examples in Chap. 5 of W. C. Reynolds, "Thermodynamics," McGraw-Hill Book Company, New York, 1968, if you want to see some additional analysis. Also see Chap. 9 of W. C. Reynolds and H. C. Perkins, "Engineering Thermodynamics," McGraw-Hill Book Company, New York, 1970.

piping leaks the slightest amount, air will be sucked in, destroying the desired low pressure. So, special provisions for removing the air will have to be provided if we elect to use water. Water has other problems at high temperatures. For example, if we want to boil water at 700°F, we will have to do it at a pressure of almost 3,100 psia! So, for a low-temperature application we would probably choose a fluid that had reasonable boiling pressures at low temperatures, and for a high-temperature application we'd select a fluid with tolerable pressures at high temperatures. This is why freons and ammonia are used in refrigeration systems and things like mercury, cesium, and rubidium are of interest for very high temperature power systems. Figure 4.1 shows the boiling temperatures of several

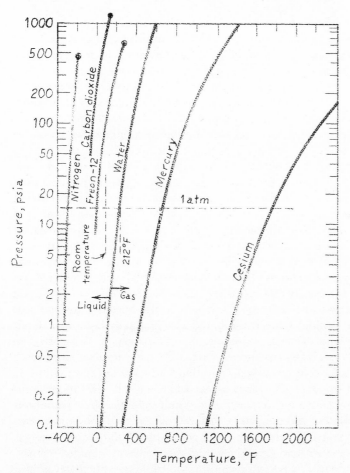

FIG. 4.1 BOILING POINTS OF SELECTED LIQUIDS.

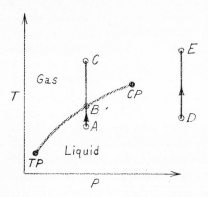

FIG. 4.2 TYPICAL *T-P* DIAGRAM FOR A FLUID.

fluids as functions of the pressure. If you have a scientific bent, check out the "Handbook of Chemistry and Physics" from your library and look up the boiling temperatures of different fluids at different pressures (often called the "vapor pressure").

Consider one of the lines on the temperature-pressure diagrams, as shown in Fig. 4.2. Points below the boiling line correspond to liquid states and points above to "vapor" or "gas" states. The lines thus mark the boundary between liquid and gas, i.e., the unique combinations of temperature and pressure for which it is possible to have liquid and gas coexisting together. For example, in Fig. 4.2 the point *A* is a liquid state. By heating the liquid at fixed pressure, we can increase its temperature to point *B*, where boiling will occur. After all the liquid has been converted to vapor, by further boiling the "state point" will move from *B* to *C*; points on the line *BC* are gas states, and point *B* is the only point where, at the fixed pressure, liquid and gas will exist together in equilibrium.

The right end of the boiling line terminates at the "critical point," "CP" in Fig. 4.2. This is the highest pressure at which a liquid-gas mixture can coexist. At pressures above the critical pressure the liquid still changes to a gas as it is heated, but not by the abrupt transition that characterizes boiling. Instead the liquid just gradually becomes thinner and thinner, and eventually disappears. This process is shown by line *DE* in Fig. 4.2, and is sometimes referred to as "supercritical vaporization."

The left end of the liquid-gas transition line terminates at the "triple point," "TP" in Fig. 4.2. Lines across which solid-liquid "phase transitions" occur (called melting or freezing) and solid-gas phase transitions (called sublimation and snowing) emanate from the triple point, as shown in Fig. 4.3. The triple point is the only point where solid, liquid, and gas can coexist together, and for a given fluid this occurs at a unique temperature and pressure. As noted previously, the triple point of water is used to define the absolute temperature scale.

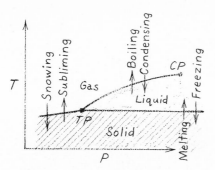

FIG. 4.3 PHASE DIAGRAM AND PHASE TRANSITIONS.

You may feel that we are getting rather technical at this point. If you are beginning to feel slightly lost, remember our objective in including this chapter—to give you just a little insight into what goes on in the minds of engineers as they are inventing and designing power systems. Struggle along and try to build as much of this insight as you can—we don't expect you to become an instant expert engineer!

In order to design a system to do a particular energy-transformation task, an engineer must have information about the "mass density" of the working fluid, usually denoted by the Greek symbol ρ ("rho"). It is a convenient fact of nature that the density of a fluid depends upon only the temperature and pressure, and hence at each point on Fig. 4.3 there is a unique value of the mass density in each phase (discontinuities occur at the lines). The density does not vary much in the liquid or solid regions; so often single numbers for each of these regions will suffice. However, the density of a gas increases markedly with increased pressure (the molecules are pushed together) and decreases with increased temperature (the molecules want to fly apart), and so charts, tables, or equations relating the gas density to temperature and pressure are always required. The charts take the form of property maps, with lines of constant density appearing as contours. Figure 4.4 shows the property map for air. For gases an algebraic equation relating the density to pressure and temperature is often used, called the "ideal gas equation" or the "perfect gas equation"

$$\rho = \frac{P}{RT} \tag{4.1}$$

where P is the absolute pressure, T is the absolute temperature, and R is a constant that varies from gas to gas. Some values for R are given in

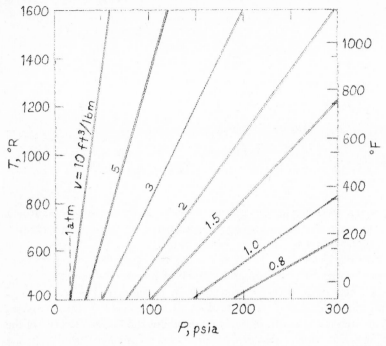

FIG. 4.4 SPECIFIC VOLUME OF AIR; THE SPECIFIC VOLUME v IS THE RECIPROCAL OF THE DENSITY.

Table 4.1. Equation (4.1) is a good approximation at points well away from the critical point, and is best at low pressures and high temperatures. For example, the density of air at 100°F (560°R) and 2 atmospheres (30 psia) is, from Eq. (4.1),

$$\rho = \frac{30 \text{ lbf/in}^2 \times 144 \text{ in}^2/\text{ft}^2}{53.3 \text{ ft-lbf/lbm-°R} \times 560°R} = 0.145 \text{ lbm/ft}^3$$

Table 4.1

Selected Gas Constants

Gas	R, ft-lbf/lbm-°R
Hydrogen	767
Helium	386
Methane	96.4
Air	53.3
Carbon dioxide	35.1

Most charts, including Fig. 4.4, give the "specific volume" (the volume per unit of mass), which is the reciprocal of the density,

$$v = \frac{1}{\rho}$$

The specific volume for air at the conditions given is

$$v = \frac{1}{0.145 \text{ lbm/ft}^3} = 6.90 \text{ ft}^3/\text{lbm}$$

Thus, at 100°F and 30 psia each lbm of air takes up a volume of 6.90 ft^3.

Flow system engineering requires data for the *internal energy* and *entropy* of the fluid. Physical chemists have determined the energy and entropy of just about every substance imaginable for practically all temperatures and pressures of interest. These are expressed in terms of the "specific internal energy" (the internal energy per unit of mass) and the "specific entropy" (the entropy per unit of mass), which are usually denoted by lowercase symbols u and s, respectively,

$$u = \frac{U}{M} \qquad\qquad\qquad (4.2a)$$

$$s = \frac{S}{M} \qquad\qquad\qquad (4.2b)$$

The dimensions of u are thus energy/mass, and hence u values are given in either Btu/lbm or perhaps cal/gram. And, the units of the specific entropy s will be either Btu/lbm-°R or cal/gram-°K. The specific internal energy and entropy depend upon the temperature and pressure, and so contours of constant u and s can be drawn to go with the contours of constant ρ. Figures 4.5 and 4.6 give examples. Then, one giant map can be drawn with all this information on it (Fig. 4.7), which can get very confusing if you don't know what's up. Good roadmaps use several colors to help tell the lines apart; engineers seldom have the luxury of good colored property maps and usually must work from a mess like Fig. 4.7, perhaps enlarged but with even more information on it.

If you are struggling to get some feeling for what this chapter is all about, you may already have made it very much worth your while. In addition to providing data on the fluid, property graphs are very helpful in explaining how a proposed energy system will work. As you read

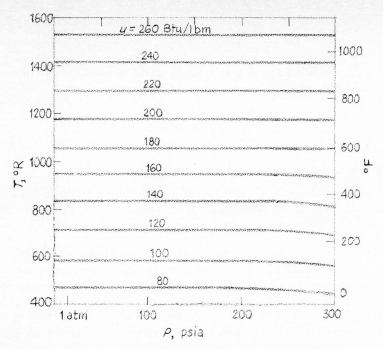

FIG. 4.5 INTERNAL ENERGY OF AIR (RELATIVE TO 1 ATM, 0°R).

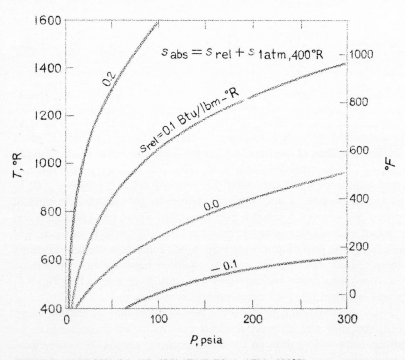

FIG. 4.6 ENTROPY OF AIR (RELATIVE TO 1 ATM, 400°R).

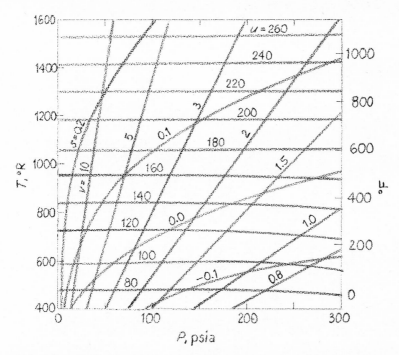

FIG. 4.7 THERMODYNAMIC PROPERTIES OF AIR (IDEAL-GAS MODEL). *s*, Btu/lbm-°R ABOVE REFERENCE OF 1 ATM, 400°R. *u*, Btu/lbm ABOVE REFERENCE OF 1 ATM, 0°R. *v*, ft³/lbm.

articles on new energy-conversion systems, you will encounter "process representations" on a variety of "thermodynamic planes," the term engineers use to describe their vast assortment of property maps. Now at least you should have some idea of what these are all about. Quite often the *temperature-entropy* (*T-s*) plane is used; this is a map with temperature as latitude and entropy as longitude, with lines of constant everything else drawn all over the place. Figure 4.8 shows the *T-s* diagram for water. Most of the lines have been removed so that you can see some white space between. An engineer would want a large graph with many more lines to do accurate calculations, and then would use a small map with just a few key lines to explain how this device works.

Figure 4.9 is a less cluttered *T-s* plane, showing only one line of constant pressure, labeled with a circle-*P*. We will use such circles to denote properties constant along lines. The bell-shaped curve is called the "vapor dome"; states inside this curve correspond to mixtures of liquid and gas (vapor); to the left of the dome is liquid and to the right is gas. The left-hand side of the dome, containing point *B*, is called the

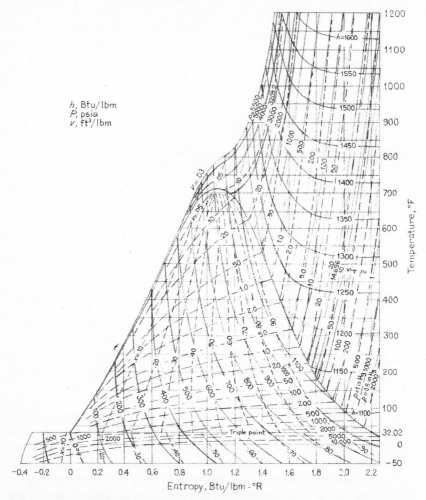

FIG. 4.8 TEMPERATURE-ENTROPY DIAGRAM FOR WATER.

"saturated-liquid line," and the right-hand side containing point *B* is the "saturated-vapor line." The top of the hill is the critical point. The solid states and the triple point are not shown. Note that the saturated-liquid line and the saturated-vapor line *coincide* on the *T-P* plane (Fig. 4.2). On the *T-P* diagram (Fig. 4.2), the vapor dome appears as the line between the triple point and the critical point, just as the face of a cliff appears as a line when the cliff is viewed from the side.

Suppose we imagine an experiment in which we heat a fluid in a balloon that permits the fluid to expand as necessary to hold the pressure

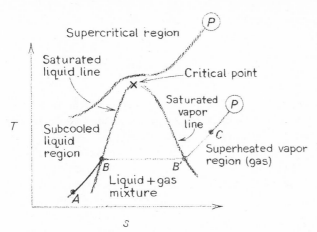

FIG. 4.9 *T-s* CHART FORM.

constant. We'll start at point *A* on Fig. 4.9, a liquid state. As we transfer energy as heat to the fluid, the temperature will of course increase. According to Eq. (3.14), the entropy will also increase. When we reach point *B* the liquid starts to boil; we will have to add more energy as heat to the fluid to evaporate the liquid completely, and again according to Eq. (3.14) the entropy of the liquid-gas mixture will increase. However, the temperature will not change as we boil off the liquid; but the mass fraction of liquid present (called the "moisture" of the mixture) will gradually go from 100 to 0% as we continue the boiling process. Equivalently, the mass fraction of vapor present (called the "quality" of the mixture) will go from 0 to 100%. At point *B'* we have evaporated all the liquid; continued heating will just warm up the gas to point *C*, with another entropy increase associated with the heating process. Alternatively, we could start at point *C* and cool the gas. This would drop the temperature and entropy (remember that heat removal reduces the entropy!). At point *B'* the liquid would start to condense, and droplets would form in the mixture. As we cool the mixture, its state would gradually change from the saturated-vapor state *B'* to the saturated-liquid state *B*, where all the fluid is liquid. Further cooling of the liquid would lower the temperature and entropy of the liquid back to state *A*. Understanding these hypothetical experiments will help you to understand the workings of energy systems, so go over this paragraph a couple of times if the ideas are not yet clear.

A little familiarity with some extra jargon will help you in reading about energy systems. If you looked closely at Fig. 4.8, you may have noticed lines of constant *h*. This is a property of the fluids called the "enthalpy" (en-thal-pee, not to be confused with en-tro-pee). The enthalpy h is defined as

$$h = u + Pv \qquad (4.3)$$

and is simply the sum of the specific internal energy and the pressure–specific volume product. (You will note that the dimensions of Pv are the same as those of u; so are the units if one takes care to convert things like ft-lbf to Btu.) It happens that the combination $u + Pv$ appears over and over again in the energy analysis of flow systems, and so engineers have gotten the physical chemists to include lines of constant h on the various thermodynamic planes. We'll see how the enthalpy arises shortly.

Another thermodynamic plane that is used mainly for analysis of refrigeration and other low-temperature systems is the *pressure-enthalpy* plane. Figure 4.10 shows a simplified P-h diagram, including a line of constant temperature. Note the locations of the liquid, gas, and mixed-phase regions. Figure 4.11 shows a more detailed P-h chart for freon-12, a common fluid used in household refrigerators. Note that portions of the constant T, s, and v lines are not shown. Still another useful plane is the enthalpy-entropy plane (h-s), shown schematically in Fig. 4.12. Again the circle-P and circle-T are used to denote the lines of constant P and T. A more detailed h-s diagram for water, as might be used by an engineer in the design of a modern steam power system, is shown in Fig. 4.13. The h-s plane is especially useful in describing systems, because h relates to the flow energy per unit mass, and s is the entropy per unit mass. Thus, vertical (h) distances relate to flow energy changes, and horizontal (s) distances relate to flow entropy changes. The h-s diagram is called a "Mollier diagram" in honor of an early engineer.

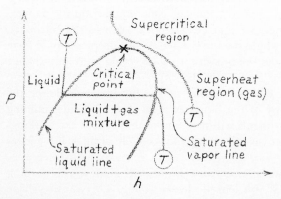

FIG. 4.10 *P-h* CHART FORM.

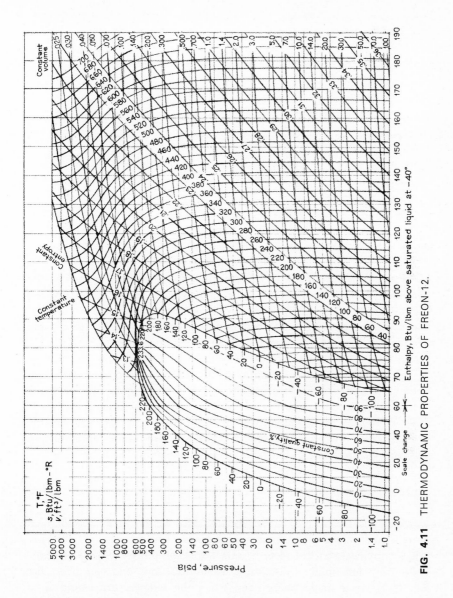

FIG. 4.11 THERMODYNAMIC PROPERTIES OF FREON-12.

We hope that you are not overwhelmed by the complexity of this fluid-property information; some appreciation for what it is and how it is used will be helpful to you in understanding an engineer's explanation of an energy system, and this can help you ask the right questions when the time comes for a public debate on the system. If you are feeling brave,

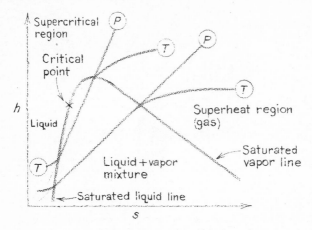

FIG. 4.12 *h-s* CHART FORM.

check out the "Handbook of Refrigeration Engineering," or look in the "International Critical Tables," or in the back of books on applied thermodynamics* to see the myriad of property charts that are available. The charts are nothing but plots of numbers, and the numbers are usually available in tables.† So, often an engineer will take his data from a table rather than from a graph.

Approximate algebraic equations that relate the internal energy and enthalpy changes to temperature changes are often useful. In general the specific internal energy can be changed by changing either the temperature or density of the fluid; but if the density is not changed by the process, then only temperature changes will produce internal energy changes. In this case, the internal energy change can be written as

$$\Delta u = c_v \, \Delta T \qquad \text{(for constant } v) \tag{4.4}$$

where the coefficient c_v is called, for historical reasons, the "specific heat at constant volume," an unfortunate name because "heat" does not appear anywhere in Eq. (4.4). If one wants to be fussy, Eq. (4.4) really holds only for small temperature differences, but one can often use Eq. (4.4) for a single phase even when the temperature change is a hundred degrees or more. Although we suggested that Eq. (4.4) holds

---

* See W. C. Reynolds, "Thermodynamics," McGraw-Hill Book Company, New York, 1968, or W. C. Reynolds and H. C. Perkins, "Engineering Thermodynamics," McGraw-Hill Book Company, New York, 1970, for example.

† See, for example, Keenan, Keyes, Hill, and Moore, "Steam Tables," John Wiley & Sons, Inc., New York, 1969, the steam power plant designer's bible!

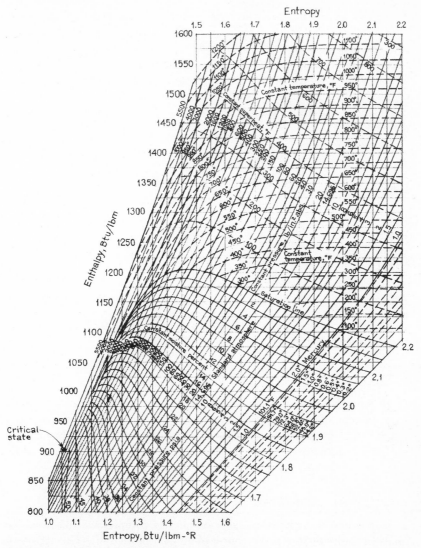

FIG. 4.13 MOLLIER DIAGRAM FOR STEAM.

only for a constant-volume process, for gases u is almost totally independent of the fluid density (except near the critical point), and so for gases Eq. (4.4) can be a good approximation for any process. Similarly, the enthalpy property depends on both temperature and pressure; but for a constant-pressure process we can write

$$\Delta h = c_p \, \Delta T \qquad \text{(for constant } p) \tag{4.5}$$

where the coefficient c_p is called the "specific heat at constant pressure."
Equation (4.5) holds for gases even if the pressure does change (except
near the critical point). The specific heats c_v and c_p are properties of the
fluid that depend upon the fluid state (temperature, pressure, and phase).
One can often get away with assuming that c_v and c_p are both "constants
of the fluid," for neither depends strongly on pressure and both increase
only slowly with increasing temperature. Typical values are given in
Table 4.2.

Table 4.2
Specific Heats of Selected Fluids at Moderate Temperatures

	c_v, Btu/lbm-°R	c_p, Btu/lbm-°R
Water*	1	1
Water†	0.336	0.446
Freon-12*	0.22	0.22
Light oil*	0.5	0.5
Mercury*	0.035	0.035
Air†	0.171	0.24
Helium†	0.75	1.25
Hydrogen†	2.43	3.42
Methane†	0.40	0.53
Carbon dioxide†	0.156	0.202

* liquid
† gas

If you have read this far without giving up, you have probably got
some inkling of the important features of working fluids, and the way
that an engineer will deal with fluid properties; good for you! Now we will
put this information together with energy balances to study some
interesting systems.

COMPONENT ENERGETICS

Any complex device contains a number of components that must be
analyzed individually in the course of an overall analysis. The approach
is:

1 Define the system.
2 Make approximations and idealizations.
3 Write the energy balance.
4 Use property information to finish the calculation.

We will now illustrate this procedure for a number of simple devices;
these examples display a systematic methodology of engineering analysis

and also serve to introduce several ideas about the systems considered. Again, our objective is not to make you skilled at such analysis, but merely to acquaint you with what is involved.

Let's first consider a tank containing 100 lbm of liquid. Suppose we want to increase the liquid temperature from 50°F to 110°F (a change of 60°R). Figure 4.14 shows the system; note the carefully positioned dots, which enclose only the liquid. Q is therefore the amount of energy transfer as heat from the immersion heater to the liquid. We will neglect energy transfer as heat from the liquid to the tank walls. Also, we will assume that at the beginning and end of the process the temperature of the liquid is uniform throughout the tank, so that each lbm has the same internal energy. This will allow us to represent the total internal energy as $U = Mu$. Then, since M is constant, $\Delta U = M \, \Delta U$. The energy balance is

$$\underset{\substack{\text{energy} \\ \text{input}}}{Q} = \underset{\substack{\text{increase in} \\ \text{energy storage}}}{\Delta U} \tag{4.6}$$

Now we use the liquid-property information to relate the internal energy change to the temperature change. Using Eq. (4.4) and a c_v value of 0.5 Btu/lbm-°R,

$$\Delta u = c_v \, \Delta T = 0.5 \text{ Btu/lbm-°R} \times 60°R = 30 \text{ Btu/lbm}$$

Substituting into the energy balance,

$$Q = M \, \Delta u = 100 \text{ lbm} \times 30 \text{ Btu/lbm} = 3,000 \text{ Btu}$$

Now the engineer (or you!) can decide how long this heating operation should take, and then determine the required heat-transfer rate. If we want to do the job in 6 min, which is 0.1 hr,

$$\dot{Q} = \frac{Q}{t} = \frac{3,000 \text{ Btu}}{0.1 \text{ hr}} = 30,000 \text{ Btu/hr}$$

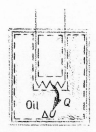

FIG. 4.14 THE SYSTEM IS THE OIL.

This is about 8.8 kw; so we could purchase a 10-kw immersion heater and expect to be able to heat the liquid easily in 6 min. However, the liquid might get too hot at the heater surface; if so, we might stir it, thereby increasing the convective heat-transfer coefficient and lowering the heater surface temperature [see Eq. (3.8)]. You can think about the energetics of the system with the stirrer in place; under what conditions would you have to take the stirrer power into consideration?

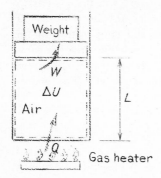

FIG. 4.15 THE SYSTEM IS THE AIR.

Let's look now at a different kind of heating process. The air in the cylinder of the little engine in Fig. 4.15 will be heated, but the piston will move in such a way as to keep the air pressure constant at 100 psia. The cylinder cross-sectional area is 2 in², and we want to move the piston such that the length L changes from 3 to 5 in as we transfer energy into the air as heat from the hot cylinder base. The initial air temperature is 500°R (40°F). The problem is to calculate the amount of energy transfer as heat required. Again we have marked the system to be analyzed by dots in Fig. 4.15. Note that, in addition to the energy inflow as heat Q, there is an energy outflow as work W done by the air on the piston. So, the energy balance is

$$\underset{\substack{\text{energy}\\\text{in}}}{Q} = \underset{\substack{\text{energy}\\\text{out}}}{W} + \underset{\substack{\text{increase in}\\\text{energy storage}}}{\Delta U} \qquad (4.7)$$

We want to solve for Q, so we'll have to come up with numbers for both W and ΔU. The work W can be calculated from the force and displacement of the piston; the force is the gas pressure P times the piston area A, so

$$W = PA\,\Delta L \qquad (4.8)$$

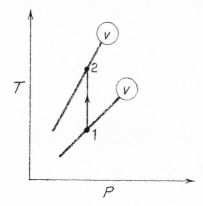

FIG. 4.16 PROCESS REPRESENTATION FOR THE AIR.

Putting in the numbers,

$$W = 100 \text{ lbf/in}^2 \times 2 \text{ in}^2 \times (5 - 3) \text{ in} = 400 \text{ in-lbf}$$

Now, since $U = Mu$ and $\Delta U = M \Delta u$, we can use Eq. (4.4) to calculate Δu if we can find the final temperature.* Now we must ponder; should we assume that the temperature remains constant? Do we have enough information to determine the final temperature? Here a process representation (Fig. 4.16) helps us to see what to do. Using Fig. 4.7 as a guide, we see that the temperature will increase as the gas volume is increased at constant pressure. If we had an accurate chart, we could read off the temperature and internal changes directly; let's instead use the ideal-gas approximation [Eq. 4.1)]. Using subscripts 1 and 2 to denote the initial and final states, Eq. (4.1) gives†

$$v_1 P_1 = RT_1$$
$$v_2 P_2 = RT_2$$

Dividing the second by the first and canceling out the equal pressures and R,

$$\frac{T_2}{T_1} = \frac{v_2}{v_1}$$

* Remember, Eq. (4.4) works for a gas even if the volume does change.
† Remember that $v = 1/\rho$.

Now, $v_2/v_1 = L_2/L_1 = 5/3 = 1.67$. So,

$$T_2 = T_1 \times \frac{L_2}{L_1} = 500°R \times 1.67 = 833°R$$

Now we can calculate Δu from Eq. (4.4),

$$\Delta u = c_v \Delta T = 0.171 \text{ Btu/lbm-}°R \times (833\text{–}500)°R = 56.9 \text{ Btu/lbm}$$

We'll need the air mass; from Eq. (4.1), the air density is

$$\rho_1 = \frac{P_1}{RT_1} = \frac{1.44 \times 10^4 \text{ lbf/ft}^2}{53.3 \text{ ft-lbf/lbm-}°R \times 500°R} = 0.54 \text{ lbm/ft}^3$$

Now, the initial volume is

$$V_1 = AL_1 = 2 \text{ in}^2 \times 3 \text{ in} = 6 \text{ in}^3 = 3.47 \times 10^{-3} \text{ ft}^3$$

so the air mass is

$$M = V_1 \rho_1 = 3.47 \times 10^{-3} \text{ ft}^3 \times 0.54 \text{ lbm/ft}^3 = 1.86 \times 10^{-3} \text{ lbm}$$

Thus, the internal energy change is

$$\Delta U = M \Delta u = 1.86 \times 10^{-3} \text{ lbm} \times 56.9 \text{ Btu/lbm} = 0.106 \text{ Btu}$$

We are now ready to use Eq. (4.7) to calculate Q. But we can't add Btu to in-lbf, so let's convert W from in-lbf to Btu;

$$W = \frac{400 \text{ in-lbf}}{12 \text{ in/ft}} \frac{1 \text{ Btu}}{778 \text{ ft-lbf}} = 0.043 \text{ Btu}$$

Finally, using Eq. (4.7),

$$Q = 0.043 \text{ Btu} + 0.106 \text{ Btu} = 0.149 \text{ Btu}$$

Note that some of the energy input as heat goes into increasing the internal energy and the rest goes out of the system as work.

Well, that was a long problem. But it was a very simple problem, and so perhaps you sense the complexity that an engineer must deal with in energy-system analysis. Shortcuts are always looked for, and there is a slightly shorter way to do this problem; noting from Eq. (4.8) that for a constant-pressure process,

$$W = P \Delta V \qquad (P = \text{constant}) \qquad (4.9)$$

and that $V = Mv$, we can write (M is constant too!)

$$W = MP \Delta v = M \Delta(Pv) \qquad (P = \text{constant}) \qquad (4.10)$$

Then, the energy balance may be written as

$$Q = M \Delta u + M \Delta(Pv) \qquad (P = \text{constant})$$

Recalling the definition of *enthalpy*, $h = u + Pv$, we see that for our constant-pressure process the energy balance is equivalent to

$$Q = M \Delta h \qquad (P = \text{constant}) \qquad (4.11)$$

So, we need only to evaluate the *enthalpy* change to get the energy input as heat for this constant-pressure process. We calculate the final temperature as before, and then,

$$\Delta h = c_p \Delta T = 0.24 \text{ Btu/lbm-}°\text{R} \times (833-500)°\text{R} = 80 \text{ Btu/lbm}$$

Finally,

$$Q = M \Delta h = 1.86 \times 10^{-3} \text{ lbm} \times 80 \text{ Btu/lbm} = 0.149 \text{ Btu}$$

as found previously. You might guess (correctly) that engineers find that the enthalpy will be particularly useful whenever one deals with a constant-pressure process.

As a third example, let's consider the heating of air in a tube. We will pass an electric current through the walls of the tube to do this job, and the problem is to calculate the amount of electrical power that must be supplied. This problem is included to illustrate the manner in which an engineer will deal with a system through which there is some fluid flow, which includes most systems of any real interest. Figure 4.17 shows the system to be analyzed. The dotted lines in Fig. 4.17a surround the region in space between the two sections 1 and 2. Mass enters at section 1 and leaves at section 2, and these mass flows carry energy in and out of our system. You might think that, if the mass flow rate is $\dot{M}$ lbm/sec, the energy inflow rate at section 1 should be $\dot{M}u$; but this is not quite correct,

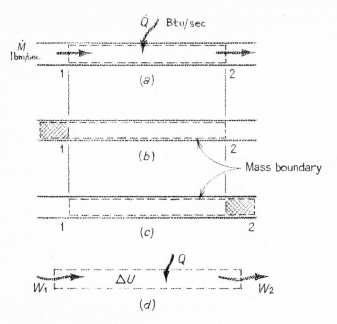

FIG. 4.17 FIGURES FOR THE HEATER ANALYSIS. (*a*) THE "CONTROL VOLUME," (*b*) THE MASS AT TIME 0, (*c*) THE MASS AT TIME *t.* (*d*) ENERGY FLOWS FOR THE MASS.

as we shall see. Since we are not sure of just how to make a proper energy balance on a system with flow entering, let's instead consider a fixed-matter system that we do know how to handle, and see what we get. This system is shown in Fig. 4.17*b*; it consists of all the material within the heater at time 0 *plus* the little piece of fluid that will enter the heater between time 0 and time *t*. Figure 4.17*b* shows the position of this fixed-matter system at time 0, and Fig. 4.17*c* shows its position at time *t*. Note that the ends of this system move, and hence there will be energy transfers as work associated with these end motions; these are the terms that we miss if we are not careful in the analysis, and they are important. The energy balance, over the time period $0 - t$, is

$$\underset{\substack{\text{energy}\\\text{in}}}{W_1} + Q = \underset{\substack{\text{energy}\\\text{out}}}{W_2} + \underset{\substack{\text{increase in}\\\text{energy storage}}}{\Delta U} \qquad (4.12)$$

We shall assume that steady-state conditions exist; this means that the amount of energy stored between sections 1 and 2, which we'll denote by U_{12}, does not change. Thus, the change in the internal energy of the fixed-matter system is

$$\Delta U = U_{\text{final}} - U_{\text{initial}}$$
$$= (U_2 + U_{12}) - (U_1 + U_{12}) = U_2 - U_1 \qquad (4.13)$$

where U_1 is the amount of energy stored in the shaded space upstream of section 1 and U_2 is the amount of energy stored in the shaded space downstream of section 2. We can relate these energy-storage terms to the mass contained in the shaded spaces,

$$U_1 = M_1 u_1 \qquad U_2 = M_2 u_2 \qquad (4.14a,b)$$

where u_1 and u_2 are the specific internal energies of the fluid at sections 1 and 2, respectively, and M_1 and M_2 are the masses of the two shaded regions. The volumes of the shaded regions can be related to the flow velocities and the flow areas at sections 1 and 2. Now a symbol problem arises; we want to use V for both velocity and volume; let's use $\mathscr{V}$ for volume and V for velocity. Then, the volumes of the shaded regions are

$$\mathscr{V}_1 = A_1 V_1 t \qquad \mathscr{V}_2 = A_2 V_2 t \qquad (4.15a,b)$$

So, since mass = density × volume,

$$M_1 = A_1 \rho_1 V_1 t \qquad M_2 = A_2 \rho_2 V_2 t \qquad (4.16a,b)$$

If we divide M_1 by t, the time it takes for the mass M_1 to cross section 1, we have the "mass flow rate" at section 1, which we shall denote by $\dot{M}_1$ (remember overdots are used to denote rates of flow, e.g., lbm/sec, Btu/sec, etc.),

$$\dot{M}_1 = A_1 \rho_1 V_1 \qquad \text{mass flow rate} \qquad (4.17)$$

Since we have steady-state conditions at each point in the tube, the principle of conservation of mass* requires that the mass coming out at section 2 must exactly balance the mass coming in,

$$\dot{M}_1 = \dot{M}_2 \qquad (4.18)$$

---

* Mass, as energy, is conserved; so we keep books on mass in the same way.

So we'll denote $\dot{M} = \dot{M}_1 = \dot{M}_2$. Then $M_1 = M_2 = \dot{M}t$, and using Eq. (4.14) in Eq. (4.13), we can now express the energy change in the fixed-matter system in terms of the mass flow rate $\dot{M}$ and the specific internal energies of the fluid at section 1 and 2,

$$\Delta U_{\substack{\text{fixed-mass} \\ \text{system}}} = \dot{M}(u_2 - u_1)t \qquad (4.19)$$

Now we have to calculate the work terms W_1 and W_2 in Eq. (4.12). The force on the upstream end of the cylinder of fluid that we are analyzing is P_1A_1, and this force pushes the fluid a distance V_1t. So, the work input to the fixed-matter system is

$$W_1 = P_1A_1V_1t \qquad (4.20a)$$

This represents the work done on the fluid mass by fluid behind it pushing it across section 1. Similarly, the work output at the other end, where the fluid pushes its way out of the pipe, is

$$W_2 = P_2A_2V_2t \qquad (4.20b)$$

This represents the work that the fluid does in pushing other fluid out of the way so that it can cross section 2. Since $v_1\rho_1 = 1$, we can multiply the right-hand side of Eq. (4.20a) by $v_1\rho_1$ without changing its value, and thus

$$W_1 = P_1A_1V_1tv_1\rho_1 = (P_1v_1)(A_1\rho_1V_1)t$$

Recalling Eq. (4.17), this is equivalent to

$$W_1 = (P_1v_1)\dot{M}t \qquad (4.21a)$$

Similarly,

$$W_2 = (P_2v_2)\dot{M}t \qquad (4.21b)$$

Now we stick Eqs. (4.21) and (4.19) back into the basic energy balance [Eq. (4.12)] and obtain

$$Q + (u_1 + P_1v_1)\dot{M}t = (u_2 + P_2v_2)\dot{M}t \qquad (4.22)$$

This is the correct energy balance for the heater tube. We worked it out for a fixed-matter system consisting of a cylinder of fluid in the act of passing through the tube; but we can interpret it as an energy balance on the system in Fig. 4.17a, the *region in space* ("control volume") between sections 1 and 2, as follows:

$$Q + (u_1 + P_1 v_1)\dot{M}t = (u_2 + P_2 v_2)\dot{M}t$$

energy
inflow
as heat

energy inflow
with mass

energy outflow
with mass

Note that the amount of energy that enters the "control volume" at section 1 with each lbm is $(u + Pv)_1$. The u is the convected internal energy per unit mass, and the product Pv is the work that must be done by the next fluid coming along to push a unit mass into the control volume. In this situation the term Pv is called the "flow work"; it will always appear in energy balances wherever mass flows across the boundaries of a system. Note that the sum $u + Pv$ is the enthalpy; in this situation the enthalpy h_1 represents the total energy flow across section 1 per unit of mass. We will use the term "flow energy" as a synonym for enthalpy,* which always appears in the energy balance on a control volume wherever mass crosses the boundaries. Now you can see why engineers are so eager to have enthalpy data on their property charts; it is a very handy property in any flow system analysis!

This was a very complicated analysis, yet it leads to a very simple final result, and this is something that is very typical of good engineering analysis. If we divide Eq. (4.22) by t, we obtain an equation for the heat-transfer rate $\dot{Q} = Q/t$,

$$\dot{Q} + \dot{M}h_1 = \dot{M}h_2 \tag{4.23}$$

rate of energy
input

rate of
energy output

Note the interpretation of Eq. (4.23) as an energy balance on the *control volume* (Fig. 4.17a), written on a "rate basis." For steady-state conditions there is no change in the energy stored within the control volume, and so no energy-storage term appears above. Rearranging Eq. (4.23),

$$\dot{Q} = \dot{M}(h_2 - h_1) \tag{4.24}$$

So, we see that the energy input to the control volume as heat shows up as an increase in the *enthalpy* of the fluid as it passes through the control volume. Simple expressions like Eq. (4.24) can be developed for many components of interest in the engineering of a power system; but what a lot of clever work it takes to get there!

---

* Sometimes the enthalpy is called the "heat content" or "sensible heat," which does not really seem very sensible to modern thermodynamicists.

In the previous piston-cylinder example the enthalpy arose in a constant-pressure process. Here it arises in a flow process, which need not be a constant-pressure process. In other words, P_2 need not be the same as P_1 for Eq. (4.24) to apply. However, it happens that the pressure drop in heater tubes like this is usually small in comparison with the value of the pressure, and so engineers often model steady-flow heater problems by assuming that the fluid pressure is constant throughout the heater.

If you are sharp, you may have noticed that this analysis carries a number of implicit assumptions. Let's make them explicit; we neglected the kinetic and potential energies of the fluid, assumed that every piece of fluid that enters comes in at the same state (which sort of ignores the effects of the tube walls on the flow), and made a similar assumption for the exit flow. These are good assumptions in most cases, and where they are not a competent engineer can fix things up by doing a more complicated theory.

Let's put in some numbers to Eq. (4.24). Suppose we want to heat 4,000 lbm/hr of air from 60 to 140°F. We use the equation-of-state information to relate the enthalpy change to the temperature change; from Eq. (4.5), which holds for gases even if the pressure is not constant,

$$h_2 - h_1 = c_p(T_2 - T_1) = 0.24 \text{ Btu/lbm-°R} \times 80°\text{R} = 19.2 \text{ Btu/lbm}$$

Then,

$$\dot{Q} = 4{,}000 \text{ lbm/hr} \times 19.2 \text{ Btu/lbm} = 7.68 \times 10^4 \text{ Btu/hr}$$

This is equivalent to 22.6 kw. A heat-transfer analysis allows the engineer to calculate the temperature distribution along the tube that will develop when this power is supplied (a careful engineer would check to be sure that the tube will not melt!).

For our next example, let's suppose we have a geothermal well that produces 10,000 lbm/hr of hot steam at 100 psia and 400°F. We would like to run this steam through a steam turbine to generate power, and are curious about how much power we could obtain. Figure 4.18a shows the system that the engineer would analyze. It is a control volume enclosing the turbine and cutting across the intake and discharge pipes at sections 1 and 2, respectively. The engineer would assume that "steady-state" conditions exist, which means that at any point within the control volume nothing changes in time (though of course the state of the fluid passing through the control volume changes as the fluid moves through). He could go through the business of taking a fixed-matter system, letting it slip a little ways through the turbine, etc., as in the previous example.

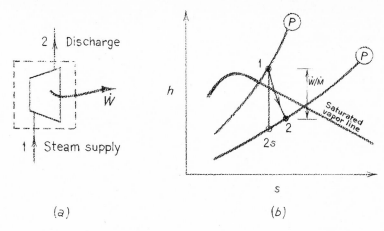

FIG. 4.18 STEAM-TURBINE ANALYSIS. (a) THE CONTROL VOLUME, (b) THE PROCESS REPRESENTATION.

You can guess what he would get; he would find that the energy input rate to the control volume at section 1 is simply $\dot{M}h_1$ and the energy output rate at section 2 is $\dot{M}h_2$ (if you are not convinced, repeat the developments of the previous example for the new system). Since he has done this numerous times as a student, he knows what he will get, and just writes down the final result*

$$\underset{\substack{\text{rate}\\ \text{of energy}\\ \text{inflow}}}{\dot{M}h_1} = \underset{\substack{\text{rate of}\\ \text{energy outflow}}}{\dot{M}h_2 + \dot{W}} \qquad (4.25)$$

Solving for the power output $\dot{W}$,

$$\dot{W} = \dot{M}(h_1 - h_2) \qquad (4.26)$$

Now, we know the mass flow rate $\dot{M}$. So, if we can establish the inlet and discharge enthalpies, we can compute the power output from Eq. (4.26). Figure 4.13 (or a larger, more complete chart) helps here. The inlet conditions are known, and we can read the inlet enthalpy off the chart; for 100 psia and 400°F, the chart gives $h_1 = 1,225$ Btu/lbm. The discharge will be to atmospheric pressure, and so point 2 must lie somewhere along the line $P = 14.7$ psia in Fig. 4.13; but where? A second-law

---

* Our reference to the engineer as a "he" should not discourage a "she" from entering the engineering profession!

analysis, which is just a tad beyond us at this point, tells us that the specific entropy of the discharge fluid must be at least as large as the specific entropy at the inlet, in order that the device may produce entropy, as the second law of thermodynamics requires. This means that the discharge state must lie to the *right* of the point marked 2s in Fig. 4.18b; but where? Wherever it lies, the difference $h_1 - h_2$ will be *less* than the difference $h_1 - h_{2s}$ (see Fig. 4.18b). So, the *maximum* power output is

$$\dot{W}_{max} = \dot{M}(h_1 - h_{2s}) \tag{4.27}$$

We can read h_{2s} from Fig. 4.13 as 1,075 Btu/lbm (drop a plumb line down from state 1 and read the intersection with the line $P = 14.7$ psia). So,

$$\dot{W}_{max} = 10{,}000 \text{ lbm/hr} \times (1{,}225 - 1{,}075) \text{ Btu/lbm}$$

$$= 1.5 \times 10^6 \text{ Btu/hr}$$

This is equivalent to 440 kw or 590 hp, enough to operate some light machinery. However, we won't be able to get the maximum power, because real turbines have friction, turbulence, and other irreversible factors that degrade the performance. Engineers rate the performance of real turbines in terms of the "isentropic efficiency," which is the ratio of the actual power output to the power output from the ideal device ("isentropic" means constant entropy; note that the ideal process $1 - 2s$ is one of constant entropy). The isentropic efficiency is usually denoted by the Greek symbol η ("eta"), with a subscript s to avoid confusion with cycle efficiency in power plant usage. It is defined by

$$\eta_s = \frac{\dot{W}}{\dot{W}_{max}} = \frac{h_1 - h_2}{h_1 - h_{2s}} \tag{4.28}$$

A modern steam turbine in this power range might have an isentropic efficiency of about 80% at "design conditions" (recall our discussion of the performance of turbines in Chap. 2). For our turbine,

$$\dot{W} = 0.8 \times 1.5 \times 10^6 \text{ Btu/hr} = 1.2 \times 10^6 \text{ Btu/hr}$$

which is about 350 kw or 470 hp. The actual discharge enthalpy h_2 can now be calculated, and point 2 fixed on the chart (see Fig. 4.18b). You should work this out and verify that $h_2 = 1,105$ Btu/lbm.

The analysis of more complex systems involves the systematic analysis of each component. Using the energy balance and property data, together with performance parameters such as the isentropic efficiency, the

Table 4.3
Steady-Flow Energy Analyses for Common Components

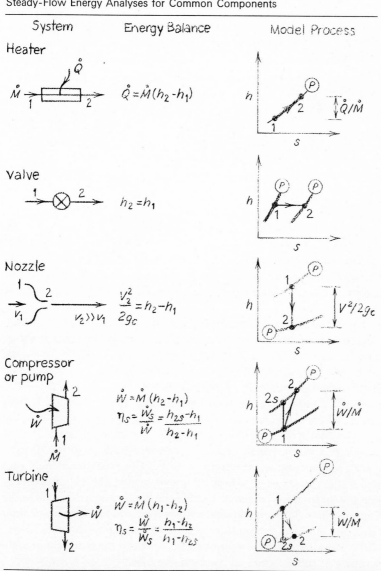

System	Energy Balance	Model Process

Heater

$$\mathring{Q} = \mathring{M}(h_2 - h_1)$$

Valve

$$h_2 = h_1$$

Nozzle

$$\frac{V_2^2}{2g_c} = h_2 - h_1$$

$$v_2 \gg v_1$$

Compressor or pump

$$\mathring{W} = \mathring{M}(h_2 - h_1)$$

$$\eta_s = \frac{\mathring{W}_s}{\mathring{W}} = \frac{h_{2s} - h_1}{h_2 - h_1}$$

Turbine

$$\mathring{W} = \mathring{M}(h_1 - h_2)$$

$$\eta_s = \frac{\mathring{W}}{\mathring{W}_s} = \frac{h_1 - h_2}{h_1 - h_{2s}}$$

engineer will work through the entire system. In Table 4.3 we have summarized the steady-flow, steady-state energy balances, process representations, and performance parameters for the most common components. You should try to derive the energy balances for yourself, listing the idealizations that are implicit in the equations. Our purpose here is not

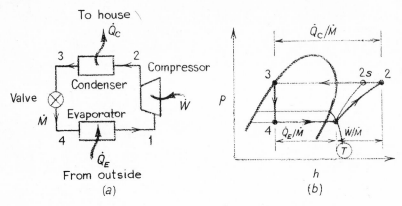

FIG. 4.19 HEAT-PUMP ANALYSIS.
(a) THE SYSTEM, (b) THE PROCESS
REPRESENTATION.

to make you an expert analyst, but you will find that the better the feeling you have for thermodynamic analysis the better you will be able to understand new energy systems. And the more you understand about technology, the better you are prepared to see that it works for the benefit of you and your society.

Let's illustrate the treatment of a full system by analysis of a heat pump. Our objective is again to give you some insight into engineering analysis, and not to give you the skill to do your own. Figure 4.19 shows the system hardware and the fluid-process representation (see Fig. 4.11). Freon-12 refrigerant enters the compressor at 10 psia and −20°F, and is compressed to 140 psia. The warm freon then passes through a heat exchanger, where it is condensed by a transfer of energy as heat from the freon to the cooler air inside the house. The liquid freon (state 3) then goes through a valve, which drops the pressure back down to 10 psia, reducing the temperature to about −35°F. The cold freon then goes through a heat exchanger, where it is evaporated (boiled) by energy transfer as heat from the colder air outside the house. Since the freon in the evaporator is at −35°F, the outside temperature can't be any lower than this, and an outdoor temperature of −10°F might be practical here. Similarly, the air in the house must be cooler than the freon condensing temperature of 105°F, which of course is the case.

The fluid is free to expand and equalize the pressure in the condenser and all the piping between the compressor and the valve. Some slight pressure drop in the direction of fluid flow is required to make the flow move and overcome friction, but this pressure drop is small compared with the pressure changes encountered in the valve and compressor.

Hence, in a preliminary analysis like this it is quite a good assumption that the pressure drops in the heat exchangers are negligible.

The energy balances on the four components are (see Table 4.3; the overdots denote rates of flow).

Compressor:

$$\dot{W} = \dot{M}(h_2 - h_1) \tag{4.29}$$

Condenser:

$$\dot{Q}_C = \dot{M}(h_2 - h_3) \tag{4.30}$$

Valve:

$$h_4 = h_3 \tag{4.31}$$

Evaporator:

$$\dot{Q}_E = \dot{M}(h_1 - h_4) \tag{4.32}$$

The game is now to pin down all four states on the property map; this will give us the enthalpies. We begin at state 1, where we know enough to fix the enthalpy; from Fig. 4.11, at 10 psia and $-20°$F,

$$h_1 = 75 \text{ Btu/lbm}$$

We know the compressor discharge pressure P_2 but do not know enough to pinpoint state 2 on the property chart. As in the turbine example just discussed, a second-law analysis tells us that the entropy of the outflow must be larger than the entropy of the inflow in order that entropy may be produced in the compressor, and hence state 2 must lie to the right of state $2s$ indicated in Fig. 4.19b. Thus, the enthalpy change $h_2 - h_1$ must be at least as large as $h_{2s} - h_1$, and hence the minimum compressor power requirement is

$$\dot{W}_{min} = \dot{M}(h_{2s} - h_1) \tag{4.33}$$

Reading from Fig. 4.11 at the intersection of the constant-entropy line through point 1 and the 140 psia pressure line, we find

$$h_{2s} = 98 \text{ Btu/lbm}$$

So,

$$\frac{\dot{W}_{min}}{\dot{M}} = 98 - 75 = 23 \text{ Btu/lbm}$$

To fix the actual state 2 we need to know the "isentropic efficiency" of the compressor, defined by

$$\eta_s = \frac{\dot{W}_{min}}{\dot{W}} \qquad (4.34)$$

Let's assume that the compressor isentropic efficiency is 85%; so

$$\frac{\dot{W}}{\dot{M}} = \frac{\dot{W}_{min}/\dot{M}}{0.85} = \frac{23}{0.85} = 27 \text{ Btu/lbm}$$

Then, from Eq. (4.29),

$$h_2 = h_1 + \frac{\dot{W}}{\dot{M}} = 75 + 27 = 102 \text{ Btu/lbm}$$

Now that we know both h_2 and P_2, we can pinpoint state 2 on the property chart.

State point 3 is on the saturated-liquid line at its intersection with the 140 psia line, and we read

$$h_3 = 31 \text{ Btu/lbm}$$

We can now calculate the condenser heat transfer per unit of mass flow; using Eq. (4.30),

$$\frac{\dot{Q}_C}{\dot{M}} = h_2 - h_3 = 102 - 31 = 71 \text{ Btu/lbm}$$

Note that the horizontal distance $h_2 - h_3$ represents condenser heat-transfer rate per unit of mass flow.

Finally, since $h_4 = h_3$ from the valve energy balance, $h_4 = 31$ Btu/lbm. Then, from Eq. (4.32),

$$\frac{\dot{Q}_E}{\dot{M}} = h_1 - h_4 = 75 - 31 = 44 \text{ Btu/lbm}$$

We can (and should) check these numbers with an overall energy balance on the entire system; this is

$$\underset{\substack{\text{energy input} \\ \text{rate}}}{\dot{W} + \dot{Q}_E} = \underset{\substack{\text{energy output} \\ \text{rate}}}{\dot{Q}_C} \qquad (4.35)$$

Dividing by $\dot{M}$, and using the numbers developed above,

$$27 \text{ Btu/lbm} + 44 \text{ Btu/lbm} \equiv 71 \text{ Btu/lbm}$$

Things check as they should.

The "coefficient of performance" of the heat pump is defined as

$$\text{COP} = \frac{\dot{Q}_C}{\dot{W}} = \frac{\dot{Q}_C/\dot{M}}{\dot{W}/\dot{M}} = \frac{71}{27} = 2.6$$

So, 2.6 times as much energy as we take from the electric grid to run the compressor will be transferred into the house as heat!

Suppose that we want to supply 100,000 Btu/hr to the house. Then, from Eq. (4.30),

$$\dot{M} = \frac{\dot{Q}_C}{h_2 - h_3} = \frac{10^5 \text{ Btu/hr}}{71 \text{ Btu/lbm}} = 1{,}410 \text{ lbm/hr}$$

So, freon will have to be circulated through the system at the rate of 1,410 lbm/hr. From Eq. (4.29), the compressor power input will have to be

$$\dot{W} = 1{,}410 \text{ lbm/hr} \times (102 - 75) \text{ Btu/lbm} = 3.8 \times 10^4 \text{ Btu/hr}$$

which is slightly more than 11 kw and about 15 hp. Note how we first calculated our way around the cycle without knowing the mass flow rate $\dot{M}$, and then found the $\dot{M}$ required to obtain the overall system performance.

You have now seen a glimpse of thermodynamic analysis. The engineer who does this for real has studied hard and long to really master the subject; he can analyze anything thrown at him with the full confidence that his systematic approach will get him a good answer. A lot of analysis like this has (or should have) been done in the study of any new energy system that you may read about. The brief glimpse here should help you understand the process of engineering analysis and perhaps will help you ask the right questions when you encounter a group trying to sell a particular energy system to the public.

POWER SYSTEMS

Fluid systems designed for the generation of power fall into two classes. The term "vapor power cycles" refers to systems in which the working fluid occurs in the system in both liquid and gas (vapor) states; steam power plants are an example. In "gas power cycles" the working fluid is a gas at every point, as, for example, in a gas-turbine engine. The term "cycle" is used because the working fluid flows around through the systems, executing a cycle of states. In "closed cycles" the fluid is completely contained with the system, while in "open cycles" the fluid comes from and is returned to the environment, which in turn closes the cycle. In this section we will discuss the basic thermodynamics of some of the more important power cycles, vapor and gas, open and closed. This will be at a level that you may well encounter in your reading; if you developed just a little insight from the previous material in this chapter, you should grasp the main ideas without difficulty.

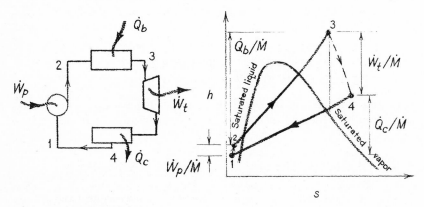

FIG. 4.20 RANKINE CYCLE.

Let's begin with a look at the "Rankine cycle," the basic cycle for vapor power systems. The system diagram and process representation are shown in Fig. 4.20. Liquid enters the pump at state 1, is compressed and fed to the boiler. The phase change occurs in the boiler, and the "superheated" steam (meaning steam hotter than the saturation temperature at that pressure) is fed to the turbine. In the turbine the flow is first accelerated to high velocity in nozzles; the high-speed flow jets against the turbine blades, which extract kinetic energy from the flow stream. The stream passes out of the turbine at a low velocity, and is condensed to return the fluid to the liquid state. As in the heat pump discussed in the previous section, the boiler and condensing processes are modeled as constant-pressure processes. The enthalpy-entropy plane (called the "Mollier diagram" in the trade) is used because it shows the energetics and irreversibilities quite well.

The energy balances (see Table 4.3) on the four components are

Pump:

$$\dot{W}_p = \dot{M}(h_2 - h_1)$$

Boiler:

$$\dot{Q}_b = \dot{M}(h_3 - h_2)$$

$$(4.36)$$

Turbine:

$$\dot{W}_t = \dot{M}(h_3 - h_4)$$

Condenser:

$$\dot{Q}_c = \dot{M}(h_4 - h_1)$$

You can see that the enthalpy differences, which are vertical distances on the Mollier diagram, represent energy transfers per unit mass of fluid passing through the device. In the boiler the fluid is heated, increasing its entropy, and in the condenser the cooling reduces the fluid entropy. A good model is that there is no heat transfer to the fluid in the pump or turbine, and hence the entropy ideally could remain constant through these devices. The increase in entropy experienced by the fluid in passing through the pump and turbine is a measure of the irreversibility of these devices caused by fluid friction, bearing friction, turbulence, etc. A glob of fluid executing the complete cycle has no total change in its entropy (it starts from point 1 and returns there), but there is an associated increase in the entropy of the environment as a result of the heat transfers Q_b and Q_c.

Let's look at some of the salient features of the Rankine cycle. Since vertical distances on the *h-s* diagram represent flow energy changes, one can get a good idea about the relative magnitudes of the works and heat from the *h-s* picture (provided that it is drawn to scale!). Note that the difference $h_2 - h_1$ is very small in comparison with $h_3 - h_4$. This means that the pump work input will be small in comparison with the turbine work output. So it won't really matter if the pump has a low efficiency; the turbine will provide much more than enough work to drive the pump. In the jargon of the trade, the Rankine cycle has a "low back work ratio," meaning that the work input to the cycle is only a small fraction of the work output. This is a main reason why vapor power cycles were successful half a century before gas power cycles, which tend to have much higher back work ratios.

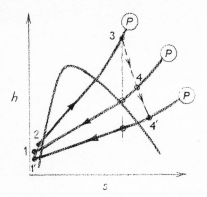

FIG. 4.21 THE PRIMED CYCLE HAS THE GREATER EFFICIENCY.

The ideas of the second law suggest that the best performance is obtained when the "heat-rejection temperature," i.e., the condensing temperature, is as low as possible. Figure 4.21 gives another view of this fact. There we see two Rankine cycles with identical boiling temperatures but different condensing temperatures. The heat input per lbm of fluid $(h_3 - h_2)$ is the same for both cycles, but the work output for the primed cycle is much greater because the turbine enthalpy change is greater. The condensing temperature must be somewhat above the environment temperature, and 100°F condensing temperatures are not uncommon.

If water is used as the working fluid, the vapor pressure at 100°F is less than 1 psia, which is less than atmospheric pressure; so, care must be taken to prevent air from leaking into the condenser, and provisions must be made for removing any air leakage and for creating the vacuum in the first place. Most steam plants employ "steam ejectors" for this purpose (Fig. 4.22). A small supply of steam bled from the boiler is

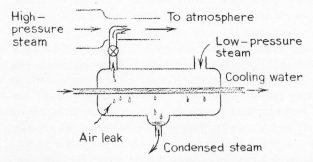

FIG. 4.22 JET EJECTORS ARE OFTEN USED TO KEEP A LOW PRESSURE IN THE CONDENSER.

passed through a "venturi" device, which accelerates the flow and drops the steam pressure. This provides a low-pressure region that can suck air (and steam) out of the condenser to provide the needed control of the pressure within the condenser.

The second law also says that raising the temperatures at which heat is added raises the cycle efficiency. You can draw a diagram like Fig. 4.21 to convince yourself of that. The peak temperature is limited by metallurgical conditions, and most steam turbines operate with inlet temperatures of the order of 1000°F.

As the steam flows through the turbine nozzles, it follows the path indicated by the dotted line 3-4 in Fig. 4.20. If this carries the steam state into the vapor dome, water droplets will begin to form, and if enough droplets form, they can do serious damage to the turbine blades by whamming against them at speeds of the order of 3,000 ft/sec. So, power plant designers must take particular care to be sure that the steam in the turbine does not get too "wet." There are many other practical problems associated with steam-turbine design, such as lubrication, speed control, and dynamical balancing, and all these factors must be carefully considered by the engineers who do the turbine design. The thermodynamicist generally has an easier time; he can simply specify that the turbine be designed for such and such an inlet state, discharging to such and such a pressure, and that the turbine should have an isentropic efficiency of so much. Then, the hard work is left up to the engineers who must design a turbine meeting these specifications.

There are many modifications to the basic Rankine cycle that are employed in large steam power plants. "Reheat," in which the turbine is split into two units and the steam leaving the first is reheated before it flows to the second, is quite common. You can draw the process representation for a reheat cycle and see how this affects the wetness of the steam in the turbine. In "regeneration," some hot steam is bled from early stages of the turbine (see Fig. 2.23), and this is in turn used to partially heat the water entering the boiler. This bootstrapping process reduces the energy that must be supplied as heat in the boiler, and thus increases the cycle efficiency. Large power stations may incorporate several stages of reheat and regeneration, each stage inching the overall system efficiency up by a percentage point or so.

At higher temperatures the vapor pressure of water becomes very high, and there is a great deal of interest in Rankine cycles using other working fluids. There was a major effort in the space program to develop Rankine power plants for space vehicles using exotic working fluids such as mercury, potassium, and rubidium, and the technological developments from these SNAP (Space Nuclear Auxiliary Power) systems are finding use in consumer energy systems. There is now a great deal of interest in "binary power cycles," systems that employ two separate Rankine cycles. A fluid well suited for high-temperature operation, such as mercury, is used in the "topping" cycle. Energy is added as heat to the mercury, and the energy removed as heat in the mercury condenser is used as the heat input to the boiler of a conventional steam power plant. A system of this sort has been in service for many years, operating with a cycle efficiency of the order of 40%. Other binary systems using things like sulfur and steam as the working fluids have been studied; these offer overall efficiencies as high as 50%, and we may see some exotic binary Rankine systems being built in the years ahead.

We have not said much about the source of energy for the Rankine cycle. In conventional systems a fossil fuel is burned in the boiler and the flames are passed over the tubes containing water. The steam rises to a "steam drum" overhead, and any liquid brought up is recirculated by gravity to the boiler. A separate superheater is usually employed to heat the steam to a higher temperature; some newer units have the superheating coils as part of the main boiler. Figure 4.23 shows a typical large power station boiler.

There are two basic schemes used in nuclear plants that use steam in a Rankine cycle. In the first, water is boiled right in the reactor to make the steam. The second uses a liquid coolant such as sodium or water at high pressure in the reactor, and then circulates the coolant to a separate heat exchanger, where energy is removed from the reactor coolant as heat to boil water in the steam-turbine cycle. Figures 4.24 and 4.25 show some of the details of these two types of nuclear power systems. We'll learn more about nuclear reactors in Chap. 7.

The "heat sink" for power plant "waste heat" (the condenser heat transfer) also deserves some comment. The second law of thermodynamics suggests that low sink temperatures give the best cycle efficiency. So, engineers try to place power plants in cold areas; a few degrees reduction in the sink temperature can mean a percent increase in efficiency, and this can mean millions of dollars over the lifetime of the plant. Rivers, lakes, and the ocean are the primary heat sinks for power stations in the United States. Water is pumped to the plant, heated by the energy removed from the steam in the condenser, and returned to the river or ocean. Temperature increases of 10 to 30°F are typical; since marine life tolerates only small variations in temperature, this type of "thermal pollution" can have a profound effect on marine life in the river, lake, or ocean near the discharge. Not only will fish die when a plant is installed, but species that come to populate the warmer waters near the plant will die when the plant is shut down for maintenance; both types

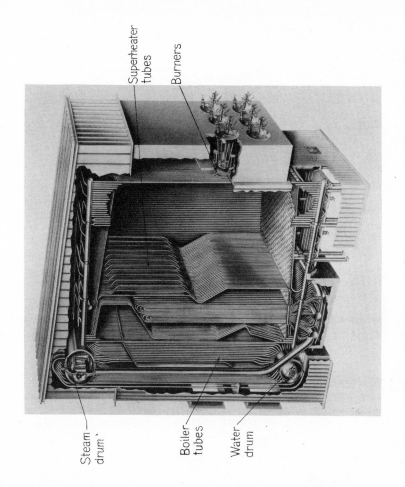

Steam drum

Boiler tubes

Water drum

Superheater tubes

Burners

FIG. 4.23 ARTIST'S CUTAWAY OF A LARGE BOILER. COURTESY OF THE BABCOCK AND WILCOX COMPANY.

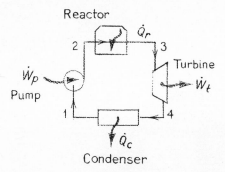

FIG. 4.24 FLOW SCHEMATIC FOR BOILING-WATER REACTOR SYSTEM.

of killing have actually occurred. Thus, there is great interest in this problem today. The impact of a new power plant on the environment must be analyzed and a report filed for public inspection, and so power plant engineers are becoming very conscious of this ecological problem.

When a water body is not available for use as the heat sink, the atmosphere must be used. "Cooling towers" such as that shown in Fig. 4.26 are often employed. Water that is used to cool the steam in the condenser is circulated to the tower, where it in turn is cooled by transfer of energy to the air before being returned to the condenser. Most cooling towers are of the "wet" variety in which the water is sprayed out into a stream of air. Some of the water evaporates into the air, removing the higher-energy water molecules from the liquid droplets and thus reducing the droplet temperature; the cooled water is then returned to the con-

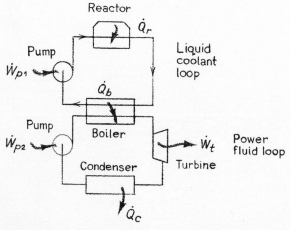

FIG. 4.25 FLOW SCHEMATIC FOR REACTORS COOLED BY LIQUIDS.

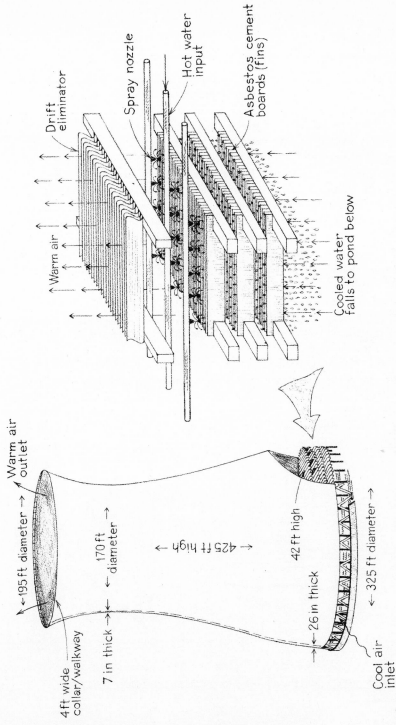

Drift
eliminator

Spray nozzle

Hot water
input

Asbestos cement
boards (fins)

Warm air

Cooled water
falls to pond below

Warm air
outlet

←195 ft diameter →

4 ft wide
collar/walkway

170 ft
diameter →

7 in thick

← 425 ft high →

42 ft high

26 in thick

325 ft diameter →

Cool air
inlet

FIG. 4.26 COOLING TOWERS FOR THE RANCHO SECO NUCLEAR POWER PLANT. (SCHEMATIC)

denser. A small amount of "makeup water" is required to replace the water that has been evaporated; about 1% of the water evaporates for each 10°F of cooling. In a 680-Mw plant the total cooling flow on a summer day might be about 15 million gallons per hour, and about 0.3 million gallons per hour of makeup water would be required. So, wet cooling towers also require a source of water for the power plant heat rejection, though normally they do use far less water than plants rejecting heat to a water stream. The water put into the atmosphere will increase the humidity in the local area, which is sometimes desirable but can make life unpleasant for residents nearby. Climatic changes are becoming factors that, like marine-life impact, must be considered by power plant engineers in selecting the cooling towers. "Dry" cooling towers do not spray water into the air. Instead the warm water from the condenser is circulated through finned tubes in the tower, as in an automobile radiator, and the energy is transferred directly to the air through the tube walls. Dry towers tend to be larger and more costly than wet towers, and have not been widely used in large power stations. The size of the tower can be reduced if the airflow is forced through by a fan, and dry towers are always of this type. Wet towers frequently use natural circulation; the beautiful hyperbolic structures of Fig. 4.26 are of this type, many of which dominate the landscape in England and western Europe. These are becoming more widely used in the United States. They are constructed from prestressed concrete shells only a few inches thick, and represent an example of high technology in the construction industry.

Let's now turn to gas power cycles, and consider in particular the gas-turbine power plant. Figure 4.27 shows the flow schematic and process

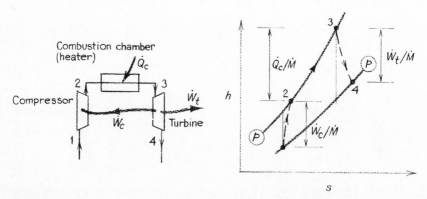

FIG. 4.27 OPEN-CYCLE GAS-TURBINE SYSTEM (BRAYTON CYCLE).

representation for the basic open "Brayton cycle"; flow enters the compressor from the atmosphere at state 1, is pressurized to state 2, heated in some manner to state 3, and then passed through the gas turbine and exhausted to the atmosphere. Energy balances on these three components give (see Table 4.3)

Compressor:

$$\dot{W}_c = \dot{M}(h_2 - h_1)$$

Heater:

$$\dot{Q} = \dot{M}(h_3 - h_2)$$

(4.37)

Turbine:

$$\dot{W}_t = \dot{M}(h_3 - h_4)$$

So, again vertical distances on the process representation (Fig. 4.27) represent the energy transfers per lbm of fluid handled by the system. Note that the compressor work input $(h_2 - h_1)$ is a large fraction of the turbine work output $(h_3 - h_4)$. The efficiency of the compressor is critical; if it is too low, then the compressor will require more work output than the turbine can supply, and the plant simply won't run! This was quite the case in the early days of gas-turbine-engine development; the compressors were so bad that the engines had to be driven externally! The first engines to be self-sustaining took practically all the turbine power to run the compressor, and so had miserable cycle efficiencies. A massive research and development effort during World War II led to improvements in compressor technology; today's compressors have isentropic efficiencies of the order of 65 to 75%, more than adequate to be matched with modern turbines with isentropic efficiencies of 70 to 80%.

Note that the Brayton cycle operates between two lines of constant pressure. On the *h-s* diagram these lines diverge to the right, and hence more power output from the turbine can be obtained if the turbine inlet temperature can be increased. The trouble is that this also increases the turbine discharge temperature, and hence a lot of energy is thrown away in the turbine exhaust. A clever scheme has been found for recovering a lot of this exhaust energy, and this development has made gas-turbine engines a significant factor in today's energy economy. The idea is to use the hot turbine exhaust gases to heat the air leaving the compressor, thereby reducing the amount of energy that must be supplied as heat to the cycle. This requires a heat exchanger, and Fig. 4.28 shows the

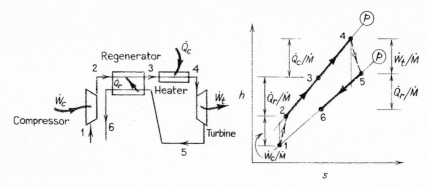

FIG. 4.28 REGENERATIVE GAS-TURBINE SYSTEM.

position of this "regenerator" or "recuperator" in the system. It is not possible to heat the compressor flow to a temperature above the turbine exhaust temperature, but with a good high-performance heat exchanger perhaps 80 to 90% of the maximum possible heating can be obtained. This greatly reduces the heat that must be supplied by fuel; cycle efficiencies of the order of 35% are possible with regeneration and good system design. This may not seem very high, but without regeneration it is difficult to do much better than 20%.

Gas-turbine engines are of great interest for mobile power sources, particularly as "prime movers" for trucks, busses, roadside construction vehicles, and even automobiles. A number of experimental automobiles powered by gas-turbine engines have been developed, trucks powered by gas-turbine engines are now on the market, and gas-turbine engines now are used in some busses. A breakthrough in regenerator design has been a big factor in this development. The old style shell-and-tube heat exchangers just are too big, too heavy for use in mobile systems, and the "periodic flow regenerator" came along to save the day. The regenerator consists of a disk of porous material, such as stainless steel or perhaps ceramic, that is rotated slowly by the engine. Hot turbine exhaust gases pass through a sector of the disk, warming the disk "matrix." The cooler compressor exhaust is passed through the other sector, taking energy out of the matrix. As the disk turns, energy is carried by the matrix from one sector to another, providing the needed coupling between the two flow streams. The matrix has an enormous amount of surface area, so that the overall volume of the regenerator is quite small. Figure 4.29 shows a typical automotive gas-turbine engine with rotary regenerator.

Gas-turbine engines are also of interest in central power stations. Cycles using helium as the working fluid have been studied for use with

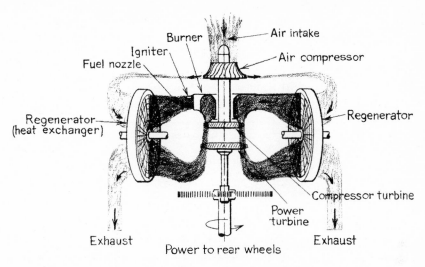

FIG. 4.29 AUTOMOTIVE GAS-TURBINE-ENGINE SCHEMATIC.

nuclear heat sources; these cycles must be closed by an additional heat exchanger that cools the flow from the regenerator discharge back down to the compressor inlet temperature so that the helium can be recycled. Such systems have also been proposed as "topping" cycles for helium-steam binary power plants, with overall efficiencies of the order of 50% being promised by the analysis. Open-cycle gas-turbine engines, using combustion of fossil fuels as the heat source, are becoming more and more important for small central power stations and for use in industrial plants that require a combination of electrical power and heat for the process (the hot engine exhaust provides an excellent source for process heat).

Aircraft jet engines are special examples of gas-turbine engines. In the basic jet engine the turbine extracts only enough energy to drive the compressor, as shown in the process diagram of Fig. 4.30. The turbine discharge, which is still quite energetic, is fed to a nozzle. The nozzle accelerates the flow to a supersonic speed; it essentially converts some of the random kinetic energy of the gas molecules into ordered kinetic energy of the flow stream. The energy balance on the nozzle can be used to derive

$$\frac{V_5{}^2}{2g_c} = h_4 - h_5 \qquad (4.38)$$

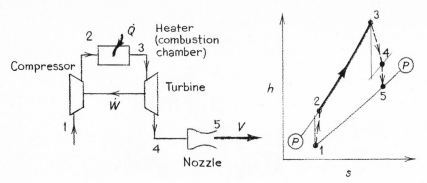

FIG. 4.30 AIRCRAFT JET-ENGINE SYSTEM.

Velocities of the order of 3,000 ft/sec can be obtained from the enthalpy differences that are available in modern engines. And the isentropic efficiency of nozzles is quite high; so practically all the available kinetic energy is realized in the nozzle exhaust. The rush of this high-speed flow from the engine produces the thrusting force. In order to produce the high-speed flow, the engine must push backward very hard on the flow; this in turn means that the flow pushes forward very hard on the engine, and this push moves the airplane forward.

Fan-jet engines are now the most common aircraft engine. These are essentially gas-turbine engines in which the net work output (turbine minus compressor work) is used to drive a separate compressor or "fan." The fan is used to blow more air, much like a propeller, and this air adds to the thrust of the engine jet. But the fan requires energy, and so not as much energy is available for the jet. In modern engines most of the thrust is obtained from the fan and a smaller amount is yielded by the jet. In effect, the fan-jet engine combines the low-weight–high-power characteristics of gas-turbine engines with the desirable thrusting properties of large fans to obtain an optimum engine for the high subsonic range.

The conventional spark-ignition engine (the "Otto" engine*), used, for example, in automobiles, is another type of gas power cycle. The pressure history in a "four-stroke" engine is shown in Fig. 4.31. With the piston near "top dead center" (TDC), the intake valve is opened and the air-fuel mixture is sucked into the cylinder during the intake stroke. The intake valve closes near "bottom dead center" (BDC), and the mixture is compressed during the compression stroke. The spark is fired shortly before the piston returns to TDC, igniting the combustible mixture. This release of energy increases the energy and pressure of the gas

* After Nikolaus Otto; you remember the ottomobile!

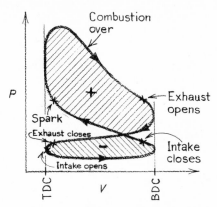

FIG. 4.31 PRESSURE CYLINDER-VOLUME DIAGRAM FOR A SPARK-IGNITION ENGINE.

molecules, and this pressure pushes the piston downward in the expansion stroke. The exhaust valve opens shortly before BDC is reached and remains open through the exhaust stroke, which returns the piston to TDC to complete the four-stroke cycle.

The work delivered to the piston on each stroke is determined by the pressure and piston displacements, and if you think carefully about it, you will realize that the work on the piston during the cycle is just the net area under the pressure-volume curve of Fig. 4.31, with the area marked "+" taken as positive and the area marked "−" taken as negative. The negative area is called the "pump loop" and represents the work that must be done to suck the flow in and push the flow out of the engine. If the spark fires too late, the positive area will be decreased and the power output of the engine will fall. Conversely, if the pump loop increases in area, the power output will also decrease. This is in fact the way that the power output is controlled. When you "step on the gas," you actually open up a throttle valve in the carburetor, which makes it easier for the engine to suck in the air (you really "step on the air"). The pressure at the intake valve is increased, and this increases the density of the fluid taken into the cylinder. The fuel flow is also increased, but the ratio of fuel to air stays about the same; the main change is in the density of the fuel-air mixture. Thus, the size of the positive area is not changed much as the load is varied, but the size of the pump loop changes markedly. The part-load efficiency of a spark-ignition engine will therefore be lowered by the throttling irreversibility, which is one reason why cars get fewer miles per gallon in the city than on a highway.

The poor part-load performance of spark-ignition engines makes them unpopular with truckers, construction companies, etc., for their vehicles

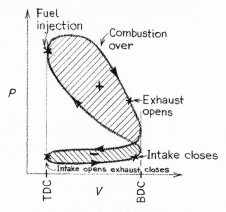

FIG. 4.32 PROCESSES IN THE CYLINDER OF A DIESEL ENGINE.

operate mainly at part load. They favor the diesel engine, which operates on a different cycle (Fig. 4.32). Air (no fuel) is sucked into the engine during the intake stroke, and squeezed on the compression stroke to a high pressure (perhaps 15 atmospheres). This greatly increases the air temperature. Fuel is then squirted into the cylinder through an "injector" and immediately begins to burn in the hot air. The burning and injection continue as the cylinder volume expands, and this tends to maintain high pressure during the power stroke. The fuel injection is terminated at a point determined by the engine load; the exhaust valve opens, and the products of combustion are pushed out, through a good muffler, we hope (and maybe someday through air-pollution-control equipment). The ability to control the air-fuel ratio gives the diesel engine better part-load performance, but the need for high pressures (to produce the high temperatures needed for ignition without a spark) means that the engine head and block must be thick and heavy. This weight is the main reason why diesel engines are not found in many automobiles.

If you are a racing fan, you may have heard of "superchargers." These are small centrifugal compressors that are used to increase the density of air before it enters a spark-ignition or diesel engine. These devices take power from the engine shaft, and do not improve the engine efficiency. However, since more flow is being handled, the *power* output of the engine can be increased, perhaps as much as 30%, with only a slight weight penalty. If you look at the diesel-engine pressures (Fig. 4.32), you see that there is still a rather high pressure in the cylinder when the exhaust valve opens. "Turbochargers" take advantage of this; the exhaust gas is fed through a small turbine, and the power output of the

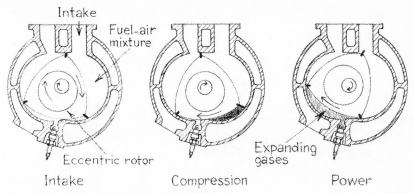

Intake

Fuel-air mixture

Eccentric rotor

Expanding gases

Intake　　　　　　Compression　　　　　　Power

FIG. 4.33 WANKEL-ENGINE SCHEMATIC.

turbine is used to drive a supercharger. In this case the efficiency is improved because the supercharger power is obtained "for free." Turbocharged diesel engines are now quite common on large construction equipment; these engines are also used to generate small amounts of electrical power (perhaps a few hundred kw) in industrial applications, emergency generators for hospitals, and a variety of other stationary applications.

A new type of spark-ignition engine appeared in the automotive world not long ago. The Wankel "rotary" engine (Fig. 4.33) uses a triangular-shaped eccentric rotor instead of conventional pistons, connecting rods, crankshafts, etc. The Wankel engine operates with intake and exhaust ports that are permanently open. As in the conventional spark-ignition engine, the intake fuel-air mixture is compressed before the spark is fired to ignite the fuel. The high pressures produced by the combustion exert a torque on the eccentric rotor and push the rotor around, exposing the exhaust port. Some of the fuel remains unburned, as in the conventional piston engine, and there is more rumor than truth in the notion that Wankel engines are "low-pollutant" engines. Its small relative size gives more room under the hood for emission clean-up devices, and this is a very strong plus for the rotary engine. But the main advantage of rotary engines is their simplicity, and it is almost certain that they will become more widely used as their technology develops.

The basic performance parameters that are important in considering engines for vehicular use are the power output per unit of weight, the fuel consumption per horsepower-hour, the efficiency (power output/fuel energy input), and the cost per horsepower of output. Values of these parameters for typical engines are given in Table 4.4.

Table 4.4

Nominal Performance Parameter of Engines in the Range of a Few Hundred Horsepower

Engine	Specific weight, lbf/hp	Specific fuel consumption, lbm fuel/hp-hr	Efficiency, %	Engine specific cost, $/hp
Spark-ignition (piston)	4	0.4	27	3
Spark-ignition (rotary)	2	0.4	30	4
Diesel	5	0.45	32	4
Gas turbine (regenerative)	2.5	0.5	30	5

PROBLEMS

To develop your skill with numbers

4.1 A household "through-flow" water-heating system heats water only as it is used. The water enters the heater pipe at 60°F. The heater power is adjusted automatically to give the desired outflow temperature; the maximum heater power is 10 kw. What is the maximum water flow (lbm/hr) that can be heated to 150°F?

4.2 A water tank contains 20 ft^3 of water at 70°F. A 20-kw electric immersion heater is used to heat the water to 140°F. How long does this take?

4.3 A house contains 15,000 ft^3 of air at a chilly 40°F. When the furnace is turned on, the heater puts energy into the air (circulated through the heater) at the rate of 100,000 Btu/hr. Assuming that all the energy remains in the air, how long does it take to warm the air to a comfortable 70°F?

4.4 A 20 lbm chunk of steel ($c_v = 0.1$ Btu/lbm-°R) is heated to 250°F, and then dumped into a pail containing 100 lbm of water at 60°F. This cools the steel and warms the water. Assuming that no energy "leaks" out of the pail, what is the final system temperature?

4.5 A 10-kw heater is used to heat each of the following from 50 to 200°F. How long does each process take?

     (a) 1 ft^3 of water ($\rho = 62.4$ lbm/ft^3)
     (b) 1 ft^3 of air at 1 atm (volume held constant)
     (c) 1 ft^3 of concrete ($\rho = 100$ lbm/ft^3, $c_v = 0.2$ Btu/lbm-°R)

4.6   What is the final pressure of the air in Prob. 4.5*b*?

4.7   The pressure of the atmosphere at an elevation of 10,000 ft is only about 10 psia. At what temperature does water boil at this elevation?

4.8   A room has the dimensions 10 ft × 20 ft × 8 ft. How many lbm of air are in the room when the temperature is 70°F?

4.9   Air in a piston-cylinder system is compressed to half its volume in a constant-pressure process. The initial volume is 0.5 ft^3, and the initial temperature and pressure are 500°R and 30 psia, respectively. The final temperature is 510°F. Calculate the amount of energy transfer as heat to or from the air for this process.

A toughie for those who dare

4.10   A large air piston system is proposed as a means for lifting heavy loads in a warehouse. The system is shown in the figure. The cylinder diameter is 10 ft, the initial length *L* is 2 ft, and the final length *L* is 12 ft. Air at 15 psia and 60°F is put into the cylinder. The air is then heated until the pressure reaches 30 psia, at which the piston floats off the lower stops. The heating continues, and the piston moves up slowly. The pressure remains essentially constant during this lifting process. When the piston comes against the upper stops, the heating is terminated and the load removed. The air is then cooled by passing water through the cooling tubes until the pressure drops to 18 psia, at which the piston starts to move down. The piston reaches the bottom of its travel, and then the cooling continues until the air temperature is returned at 60°F. Calculate the weight of the piston, the weight of the load, the work done by the piston on the load, the work done by the air on the piston for each part of the process, the amounts of energy transfer as heat for each part of the process, and the cycle efficiency (useful work output/heat-energy input). *Hint:* If you don't draw the process representation to get your thinking started and if you aren't careful in your definition of the system, you probably will go mad.

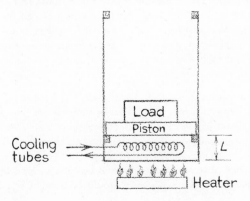

Some more interesting analyses

4.11   A 300-Mw power plant is to be located on a river where the water volume flow averages 800 ft^3/sec. A nearly extinct species of fish live in the river; biologists report that these fish will die if the water temperature is increased by more than 10°F by the power plant "waste heat" rejection.

Estimate the temperature rise; if there is a problem, discuss possible alternatives; if there is not, discuss the simplifications used in your analysis that might be too optimistic for the fish, and if possible improve your estimate. Is it worthwhile to worry about the fish at all?

4.12   A small lake covers 1 square mile with an average depth of 10 ft. A 1,000-Mw nearby power station is having trouble with its cooling towers, and wants to use the water in the lake for cooling during a 48-hr period in which they will repair the cooling towers. A conservationist group opposes the plan on the grounds that the heating of the lake water would kill a rare species of fish that live only in this lake; a biologist estimates that 75% of the fish would survive a 5°F increase in the water temperature. Analyze this problem quantitatively to decide if there is a real danger to the fish. Estimate the dollar loss to the utility company if they are forced to shut down the power station during the cooling-tower repairs. Discuss the situation. Is the conservationists' problem real or imagined? What values are in conflict? How might the situation be resolved? What would you do if you were the utility company president? What compromises might be sought?

4.13   It takes about 144 Btu to melt 1 lbm of ice. Suppose that all the "waste heat" rejected into the atmosphere by United States power plants eventually goes to melting polar ice. Estimate the amount of ice that is melted each year and the annual increase in ocean depth. Discuss the implications of your numbers.

4.14   In order to ease the thermal-pollution problem, it is proposed that power plants reject heat by nighttime radiation to deep space. Nuclear energy might be used as the heat source, and the "waste heat" would be put into lake water during the day. The lake water would then be pumped through the radiators at night. Suppose 100 Btu/hr could be rejected from each square foot during the 12-hr nighttime period. Estimate the radiator surface area required for a 1,000-Mw plant. Discuss the social, economic, and political problems that would emerge in discussions of this proposal.

4.15   A proposed scheme for generation of electrical energy involves exploding a nuclear bomb underground to create a large reservoir of hot, high-pressure steam. The steam would then be mined for use in a once-through steam power plant, condensed, and then returned to an expired cavity for storage. A bomb of a few "megatons" would release of the order of 10^{13} watt-hours. Perhaps half of this thermal energy could be captured by the power plant, and the power plant efficiency would be about 25%. How often would a 1,000-Mw power plant have to be "refueled"? Discuss the technical rationality of this proposal, and the social, political, and human values that bear on the proposal.

For your independent evaluation

4.16 Read the articles below on solar energy technology, and write a brief account of the state-of-the-art. Discuss the incentives for development of this technology; when do you think we could reach the point where half of our electrical energy is produced from solar sources?

(a) Solar Energy Progress, *Mechanical Engineering*, July 1970, p. 28
(b) Solar Powered Refrigeration, *Mechanical Engineering*, June 1971, p. 22
(c) Power in the Year 2001, Part 3, Solar Power, *Mechanical Engineering*, November 1971, p. 33
(d) Solar Sea Power, *Physics Today*, January 1973, p. 48

4.17 Read one of the following articles, and write a brief description of the system in your own language. Discuss the technological development that is required, and the values questions that impact on this technology.

(a) Air Storage Power Plant, *Mechanical Engineering*, November 1970, p. 29
(b) Combined Helium and Steam Cycle for Nuclear Power Plants, *Mechanical Engineering*, August 1971, p. 14
(c) Power in the Year 2001, Part 2, Thermal Sea Power, *Mechanical Engineering*, October 1971, p. 21
(d) Solar Sea Power, *Physics Today*, January 1973, p. 48

4.18 The systems described in the following articles have been developed to a considerable degree. Read these papers, and then discuss the relative outlook for future development of each.

(a) The Gas Turbine—Its Growth in the Electric Utility Industry, *Mechanical Engineering*, October 1970, p. 34
(b) Russia's 1000 Mw Gas Turbine, *Mechanical Engineering*, December 1970, p. 26
(c) Geothermal Power, *Scientific American*, January 1972, p. 70
(d) Steam Turbines, *Scientific American*, April 1969, p. 100
(e) Gas Turbines: An Industry with Worldwide Impact, *Mechanical Engineering*, March 1973, p. 33

4.19 The articles below describe old and new technologies applicable in thermal systems. Read these, and then discuss the ways in which the technology of each is likely to move ahead and the possible impact of each upon power systems.

    (*a*)   Cooling Towers, *Scientific American*, May 1971, p. 70

    (*b*)   The Heat Pipe, *Scientific American*, May 1968, p. 38

    (*c*)   Rotary Engines, *Scientific American*, February 1969, p. 90

    (*d*)   The Stirling Refrigeration Cycle, *Scientific American*, April 1965, p. 119

4.20 Read the following articles describing some of the history of internal-combustion engines. What values were operative for the early developers of this technology? What other values are important today as motivation for continued development of this technology? Write a short paper treating the impact of the technology and values of the early pioneers on our present society.

    (*a*)   The Origin of the Automobile Engine, *Scientific American*, March 1967, p. 102

    (*b*)   Rudolph Diesel and His Rational Engine, *Scientific American*, August 1969, p. 108

Translating down and up

4.21 Look through some recent issues of *Mechanical Engineering*. Pick an article that deals with a fluid energy system that interests you, and write a brief article about the system that would explain it simply to a class of high-school science students.

4.22 Look through some recent issues of *Scientific American*. Find an article dealing with a fluid energy system. Write a brief description of how the system works, using the terminology and approach presented here and expanding on what was discussed in the article.

5

MAKING CHEMICAL
BONDS WORK

In which we tap the energy
of molecular marriages

MOLECULAR STRUCTURE AND CHANGE

Many important energy technologies involve chemical changes of some
sort. Your understanding of these technologies will be greatly enhanced
if you can develop a little understanding of the microscopic structure of
matter. Physicists are today engaged in an exciting game of detective
work trying to discover the truly fundamental particles of nature and to
organize this information in some coherent way. Just as the universe has
galaxies, solar systems, planets, and moons, matter appears to have many
echelons of organization. The largest building block of matter that we
call "microscopic" is the "molecule"; molecules in turn are collections

of "atoms"; atoms in turn are built up from electrons, protons, and neutrons. And in turn these are built up from . . . well, this is what the physicists are trying to decide today. In spite of the fact that physicists do not yet fully understand the fundamental particles of matter, what is known does provide a rather good picture of atoms and molecules, the way that atoms combine to make molecules, and the way that molecules interact to form other molecules. The science of all this is called "chemistry"; it is one of the fundamental tools used by engineers in the design of many sorts of energy systems. This chapter provides a brief introduction to some aspects of chemistry, with applications to a variety of energy systems. Our objective is not to make you an overnight chemist, but rather to acquaint you with the basic ideas and terminology, and the way that they are used in engineering analysis of energy systems.

Let's start with one of the useful models for atoms; you probably know that atoms have a center called the "nucleus" around which tiny things called "electrons" whirl and swirl something like planets around the sun.* An electron has a property called "negative electric charge" that we will learn more about in the next chapter. The nucleus (atom center) is composed of tightly packed particles called "protons" and "neutrons." Each proton has a "positive electric charge" equal in magnitude to the electric charge of an electron; neutrons have no electric charge. Each atom contains the same number of protons as electrons, and thus the total electric charge of any atom is zero. Since electric charges of the same sign repel one another, and electric charges of opposite sign attract one another, the negatively charged electrons are pushed away from one another and pulled toward the positively charged protons in the nucleus. Only their kinetic energy prevents them from falling into the nucleus. The protons in the nucleus, all being similarly charged, tend to push away from one another. The fact that they remain tightly clustered in spite of the powerful repulsive electric forces between them indicates that there must be even stronger cohesive forces in the nucleus. These stronger forces give rise to nuclear energy, which we shall discuss in Chap. 7. The masses of the individual particles in an atom are

    Electron 9.1083×10^{-31} kg

    Proton 1.67239×10^{-27} kg

    Neutron 1.67470×10^{-27} kg

---

* Planets orbit on a plane; electrons apparently change their orbital plane continuously, and thus wander randomly about on a sphere.

Note that the proton and neutron have about the same mass, and that the mass of the electron is much less. Figure 5.1 shows some representations of a few important atoms.

Atoms are differentiated by the number of electrons. The electron orbits appear to be grouped in "shells," each of which can contain a particular maximum number of electrons. The inner shells are filled in more complex atoms, but the outer shells may or may not be filled to capacity. For example, each carbon atom contains six electrons (two filling the inner shell, and four in the second shell) having a total electric charge $-6e$ (the symbol e denotes the *magnitude* of the electric charge of an

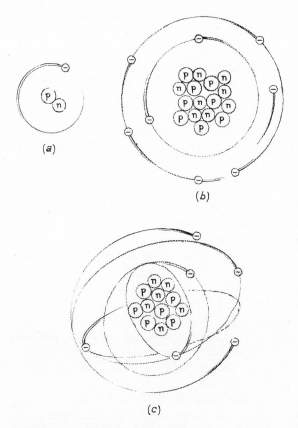

(a)

(b)

(c)

FIG. 5.1 SCHEMATICS OF THREE IMPORTANT ATOMS. p PROTON, n NEUTRON, — ELECTRON. (a) HYDROGEN ATOM, (b) OXYGEN ATOM, (c) CARBON ATOM.

electron) and six protons with a total electric charge of $+6e$, giving a total charge of $-6e + 6e = 0e$. The carbon nucleus also has six neutrons. The atoms are ordered according to the number of electrons (or protons); this is called the "atomic number" and is usually denoted by Z.

The letters are only part of the chemical designation of the elements. Frequently fore-subscripts and aft-superscripts are also used. For example,

$$_6C^{12} \quad _1H^1 \quad _1H^2 \quad _{92}U^{235}$$

In each case the letter identifies the element, the fore-subscript is the atomic number (number of electrons or protons), and the aft-superscript is the total number of particles in the nucleus (protons plus neutrons). All carbon atoms have six electrons, and so the fore-subscript is really redundant to the letter. However, physicists like to use the fore-subscript as a check in electron counting, so you will see it included sometimes and left out others. Did you notice that $_1H^1$ and $_1H^2$ are both some form of hydrogen? We'll discuss this shortly.

Each atom has a mass determined by its composition. A special mass scale called the "atomic weight" is used to compare atoms; on this scale the carbon atom $_6C^{12}$ by definition has an atomic weight of exactly 12 amu ("atomic mass units"). Mercury atoms have an average atomic weight of 200.61, and hence a mass $200.61/12 = 16.72$ times that of the $_6C^{12}$ atom.

Atoms combine to form "molecules," assemblies of two or more atoms. The electrons in the outer shell of each atom are "shared" by the atoms; you can think of these shared electrons as orbiting around all the atomic nuclei in the molecule, as shown in Fig. 5.2, and this is the "glue" that holds the molecule together. The energy associated with this glue is called the "chemical bond" or "binding" energy. The binding energy depends upon the atoms that are joined and the way that they share electrons. Chemical reactions occur when atoms combine to form molecules or when molecules combine to form other molecules. Atoms that have outer shells filled to capacity with electrons have no room to spare and hence are unable to combine to form molecules; these are the "inert" elements *helim, neon, argon, krypton, xenon,* and *radon.*

Molecules also are electrically neutral; i.e., they contain the same number of protons as electrons. Sometimes atoms or molecules lose or gain one or more electrons, and then of course they have a net positive or negative electrical charge. In such a state they are called "ions." Since particles with a charge can be pushed about by electric fields, as we will discuss in Chap. 6, ions can be pushed about by voltages. This

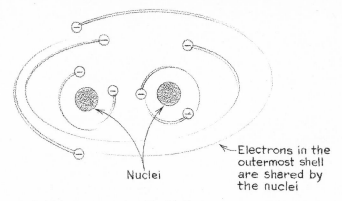

Nuclei

Electrons in the
outermost shell
are shared by
the nuclei

FIG. 5.2 THE IDEA OF MOLECULAR GLUE.

behavior is basic to electrostatic precipitators, an important type of
air-pollution-control device. Conversely, if ions are carried by material,
they in effect form an electric current, and this too has many practical
applications.

We mentioned earlier that there is more than one kind of hydrogen
atom. Atoms that contain the same number of protons but different
numbers of neutrons are called "isotopes" of the same element. Since
only the charged particles are involved in gluing atoms into molecules,
the number of neutrons (remember that neutron particles have no
electrical charge) does not matter as far as chemistry is concerned, and
hence the isotopes of a given element all have the same chemistry. In
nature there is a fixed percentage of each isotope of each element, and the
atomic weights tabulated in "periodic tables of the elements" refer to the
average as found on earth. For example, the atomic weight of uranium
is 237.9. The isotope of interest in nuclear reactors is $_{92}U^{238}$; natural
uranium is mostly $_{92}U^{235}$. Figure 5.3 shows the isotopes of hydrogen
pictorially. Physicists have created many different isotopes of every

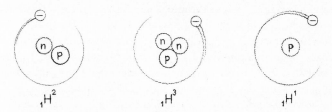

FIG. 5.3 ISOTOPES OF HYDROGEN.

substance; most of these isotopes do not last for very long but change to more common isotopes by processes of radioactive decay. We'll learn more about this in Chap. 7.

Molecules are represented symbolically in terms of the atoms that are in them. For example, the symbols C and O stand for "carbon" and "oxygen," respectively, and CO represents the "carbon monoxide" molecule (one carbon atom and one oxygen atom), that poisonous product of internal-combustion engines. And CO_2 represents "carbon dioxide" (one carbon atom and two oxygen atoms), the gas that plants breathe. Normally just the letters are used, and aft-subscripts are used to denote the number of atoms of each type in the molecule.

Atoms sometimes gain or lose one or more electrons, and then they are called "ions." Ions are represented by letter symbols with superscripts that show the charge. For example, Ag^+ represents an ion of silver having a charge of $+1e$ (one electron is absent), I^- represents an ion of iodine with charge $-1e$ (one electron added), and Al^{+++} is an aluminum ion with charge $+3e$ (with three electrons missing).

Chemists write chemical equations using these symbols. For example,

$$CH_4 + 2O_2 \rightarrow 2H_2O + CO_2 \qquad\qquad (5.1)$$

is an equation for the reaction of methane (CH_4) with oxygen (O_2). The methane molecule consists of one carbon atom and four hydrogen atoms, and the oxygen molecule consists of two oxygen atoms. One CH_4 molecule and two O_2 molecules react to form two molecules of water (H_2O) and one molecule of CO_2. Note that there are the same number of C atoms (one) on both sides of the equation, the same number of O atoms (four) on both sides, and the same number of hydrogen atoms (four) on both sides. Chemical equations *always* must be consistent with the principle of "conservation of atoms," which says that in the absence of nuclear reactions (Chap. 7) *atoms cannot be created or destroyed.*

Chemists sometimes interpret equations like (5.1) in another way. They use a material unit called the "mole," which is a mass numerically equal to the atomic weight. For example, one "gram mole" (often abbreviated gmole or just plain mole) of C^{12} will be 12 grams. If you think about it, you will see that Eq. (5.1) can be interpreted as saying that one gmole of CH_4 reacts with two gmoles of O_2 to form two gmoles of H_2O and one gmole of CO_2. This is because the "atomic weight" is really a self-consistent mass measure for the various atoms. Sometimes the term "lbmole" (pound-mole) is used; this is a mass in lbm numerically equal to the atomic weight. Equation (5.1) can be interpreted as one lbmole of CH_4 reacting with two lbmoles of O_2 to form two lbmole of H_2O and one lbmole of CO_2. Chemists are fond of the mole, and will often state the amount of chemical involved in some process in terms of moles (usually meaning gram moles). In Eq. (5.1) there happens to be three moles on both sides of the equation. But there are other reactions in which the number of moles change; Eq. (5.3) is an important example.

Some chemical reactions result in formation of ions. For example,

$$Ag \rightarrow Ag^+ + e^- \tag{5.2}$$

describes the ionization reaction of silver, in which a silver atom (Ag) loses an electron (e^-) in forming the silver ion (Ag^+).

There are jillions of possible chemical reactions; perhaps you were discouraged about chemistry at some early age by the thought of having to memorize all nature's possible reactions. But there is a science to chemical reactions, and the trained chemist can forecast all the possible reactions between given chemicals using some fundamental ideas like energy, entropy, and conservation of charge. So, it is not really such a big mystery after all.

REACTION THERMODYNAMICS

Let's now consider the energetics of chemical reaction. Suppose we dump some hydrogen and oxygen into a tank and allow them to react to form water. The chemical equation is

$$2H_2 + O_2 \rightarrow 2H_2O \tag{5.3}$$

This says that two molecules of hydrogen will react with one molecule of oxygen to form two molecules of water.* Recall that the "bonding energy" is the energy associated with the glue (shared electrons) that hold a molecule together. Now, the bonding energy of the two H_2 molecules plus the bonding energy of the O_2 molecule is *more* than the bonding energy of the two water molecules. So, when the reaction occurs, this extra energy is "released" and goes to some other form. In combustion reactions this extra bonding energy is immediately converted into disorganized translational, rotational, and vibrational molecular energy (i.e., internal energy); if the reaction chamber is rigid and insulated, so that none of the energy can escape as work or heat, the result is a dramatic increase in the temperature of the gas in the chamber. This conversion of

---

* Note that the left side has three moles and the right side only two. *Mass* is conserved, but *molal mass* is not.

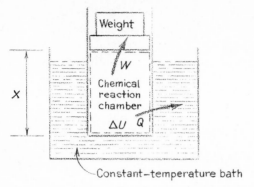

FIG. 5.4 REACTIONS CAN BE HELD AT CONSTANT PRESSURE AND TEMPERATURE.

chemical bonding energy to internal energy is sometimes loosely called "chemical heating."

In principle it is possible to maintain fixed values for both the temperature and the pressure of the gases during a combustion reaction. We can imagine using a piston-cylinder system as the reaction chamber, and allowing the piston to move to maintain a fixed pressure in the cylinder. Then, we imagine cooling the mixture to maintain the contents at a fixed temperature. Information about chemical reactions is usually given for this hypothetical reaction at a particular set of temperature and pressures (usually 1 atmosphere and 298°K). Let's look at the energy balance to see what information can be supplied (Fig. 5.4). We allow the reaction to occur and simultaneously do the expansion and cooling required to maintain fixed temperature and pressure. The energy balance on the reaction chamber is

$$0 = \Delta U + Q + W \tag{5.4}$$

energy input   change in energy storage   energy output

Here U represents the *total* internal energy of the stuff in the cylinder, *including* the chemical bonding energies of whatever molecules are there. The work term is expressible in terms of the pressure P, piston area A, and piston displacement Δx; the force is PA; so the work is

$$W = PA\,\Delta x$$

Since the cylinder volume is simply AX, this is equivalent to

$$W = P\,\Delta V$$

where V is the volume of the cylinder. But since the pressure is constant, this is also the same as

$$W = \Delta(PV) \tag{5.5}$$

Substituting Eq. (5.5) into Eq. (5.4), and solving for Q,

$$Q = -\Delta U - \Delta(PV) = -\Delta(U + PV)$$

Q could be either positive or negative, depending on the change in $U + PV$. Now, we define the "total enthalpy" of the contents of the cylinder by

$$H = U + PV \tag{5.6}$$

So, the amount of heat that must be removed to maintain the temperature and pressure constant is

$$Q = -\Delta H = -(H_{final} - H_{initial}) \tag{5.7}$$

The initial condition has hydrogen and oxygen in the cylinder; these are called the "reactants." The final condition has just water in the cylinder; this is called the "product" of the reaction. When chemists want to indicate that the reaction has been carried out at the "standard reference conditions" of 1 atm and 298°K, they use a little superscript zero. So, for a constant-temperature–constant-pressure reaction at the standard reference conditions, they would write

$$Q = -\Delta H^0 = -(H^0_{products} - H^0_{reactants}) \tag{5.8}$$

The quantity ΔH^0, which is the change in the enthalpy that occurs when the reaction takes place at the standard reference conditions, is called the "enthalpy of combustion." The negative of ΔH^0, which from Eq. (5.8) is the same as Q, is called the "heat of formation" or the "heat of reaction." Values of ΔH^0 for thousands of reactions have been measured or worked out from theory and are tabulated in reference books like the "Handbook of Chemistry and Physics." Sometimes they are given for one mole of fuel (in this case H_2) and sometimes for one mole of reactants (in this case $\frac{2}{3}H_2 + \frac{1}{3}O_2$); sometimes $+\Delta H^0$ is tabulated and other times $-\Delta H^0$ is tabulated; so one must read the table captions carefully.

In reference books on chemical properties one finds tables giving chemical thermodynamic properties of various compounds. Included in these tabulations are the "enthalpy of formation" or "heat of formation" (ΔH^0 or Q), which refer to simple chemical reactions in which the compound is formed from its basic elements. For example, the enthalpy of formation of H_2O (gaseous water) is listed in the "Handbook of Chemistry and Physics" as $\Delta H^0 = -57.8 \times 10^3$ cal/gmole, and the heat of formation is given as $Q = +57.8 \times 10^3$ cal/gmole [check Eq. (5.8)]. Also given in these tables is the specific entropy of the compound at the standard reference state (1 atm, 298°K), denoted by S^0; for H_2O (vapor) $S^0 = 45.11$ cal/gmole-°K. The table lists another quantity called the "free energy" of formation ΔG^0, which is defined by

$$\Delta G^0 = (H - TS)_{\text{products}} - (H - TS)_{\text{reactants}} \tag{5.9}$$

If you recall the relationship between the entropy change, absolute temperature, and energy transfer as heat for a reversible process [Eq. (3.14)] and use this in Eq. (5.4), you can show that the work done during the constant-pressure–constant-temperature reaction is

$$W = -\Delta G^0$$

So $-\Delta G^0$ represents the maximum energy that could be derived as work as a result of the chemical reaction. For water $-\Delta G^0 = 54.64$ cal/gmole. The "fuel cell," which we will discuss shortly, is a device that attempts to capture this available ("free") energy.

Let's now suppose we dump a bunch of chemicals into an insulated constant-volume pot, allow them to react to form other chemicals, and then want to know the composition of the final mess. For example, if we use methane (CH_4) and air (mainly O_2 and N_2), many different types of molecules will be formed, including H_2O, CO_2, CO, NO_2, NO, and some O_2, N_2, and CH_4 may remain. Can you guess what scientific principles are used to calculate how much of each chemical will appear in the final mixture? Certainly conservation of atomic species is required; we have O, N, C, and H atoms; so this gives us four equations. The energy balance is a fifth equation. But we have many more unknowns (the final temperature, pressure, and the amount of each chemical present in the mixture, at least nine as given above). Obviously more information is needed, and this is provided by the second law of thermodynamics. If the reaction chamber is insulated and of fixed volume, so that no energy can leave the system as heat or work, the second law requires that the entropy can't go down. So, as reactions occur and change the mixture composition, each reaction can only increase the entropy. The final equilibrium condition will have to be the state of *largest possible entropy*.

The physical chemist or chemical engineer who wants to calculate the final mixture plays a mathematical game in which one adjusts the temperature, pressure, and assumed composition of the mixture, each time being sure that atoms and energy are conserved; one must calculate the entropy of each trial mixture until the mixture with the largest possible entropy has been found. This last solution gives the equilibrium composition. The "equilibrium constant" that you may see in tables of chemical thermodynamic properties is a tool that helps in this sort of calculation. Nowadays such calculations are done swiftly and routinely by computers, and such calculations are an important part of the engineering analysis behind any combustion system.

COMBUSTION TECHNOLOGY

The reaction process just described was a quite hypothetical "batch" process. More often one is interested in reactions in a steady-state system. For example, in combustion chambers air and fuel flow continuously in, the reaction occurs as the flow goes along, and the hot products of combustion continuously flow out. Figure 5.5 shows a schematic of this process. It is usually reasonable to assume that there is no energy transfer as heat from the chamber and that the process occurs at constant pressure. Hence, the energy balance on the control volume shown is

$$\dot{M}_{air}h_{air} + \dot{M}_{fuel}h_{fuel} = \dot{M}_{prod}h_{prod}$$

where the $\dot{M}$ terms are the mass flow rates (e.g., grams/sec) and the h terms are the enthalpies (e.g., cal/gram). The enthalpy of the products depends upon the composition of the exhaust gases (the amounts of H_2O, CO_2, etc.); thus, a second-law analysis must be involved in the calculation of the composition and temperature of the product gases. Since the combustion chamber is insulated, there is no energy transfer as heat. Such a process is called "adiabatic," and the product temperature is called the "adiabatic flame temperature." The combustion-chamber engineer must be sure to prescribe the proper air and fuel flows to give the desired adiabatic flame temperature, and thermodynamics is the science that he uses to make this calculation.

FIG. 5.5 AN ADIABATIC COMBUSTION CHAMBER.

The adiabatic flame temperature can be controlled by adjusting the air/fuel ratio. The mixture that supplies just the proper amount of oxygen to combust all the fuel is termed "stoichiometric." If there is a surplus of oxygen, the mixture is "lean," and if there is a surplus of fuel, the mixture is "rich." Internal-combustion engines tend to operate slightly rich, and so not all the fuel is combusted; this gives rise to air pollution by "unburned hydrocarbons." In contrast, gas-turbine engines operate quite lean, and hence present practically no problems with unburned hydrocarbons. The highest adiabatic flame temperatures are produced with nearly stoichiometric mixtures, and so the lean-running gas turbine operates with a much cooler flame temperature.

The equilibrium composition of the combustion products depends on the air/fuel ratio, which also regulates the adiabatic flame temperature. At high temperatures molecules tend to "dissociate" (break down into individual atoms), and at still higher temperatures the atoms tend to ionize, thereby making the gas capable of conducting electricity. This has important applications in magnetohydrodynamic (MHD) power generators, as we shall discuss in the next chapter. Figure 5.6 shows the adiabatic flame temperature and equilibrium mixture composition for combustion of methane gas (CH_4) with air. Note that the carbon monoxide content increases dramatically with increased temperature. Thus, spark-ignition engines tend to produce a large amount of CO during combustion. Eventually the product mixture is cooled down to ambient temperatures, and thus you might think from Fig. 5.6 that the CO level would go down as the gas cools. However, most of the cooling occurs rapidly in a "quenching" process as the gas contacts the relatively cold cylinder walls. At low temperatures the rate of approach to equilibrium is very slow, and so the CO composition is essentially "frozen" by the quenching process at something close to the equilibrium composition in the high-temperature flame. Automotive engineers have been working on this quenching problem for some time, trying to allow the mixture to exist at moderate temperatures for enough time for the CO fraction to "relax" back to the relatively low levels that occur in equilibrium mixtures at moderate temperature before the final quenching to ambient temperatures. Indeed, improved exhaust-valve and manifold designs have led to reductions in the CO emissions from spark-ignition engines, but they remain much worse in this respect than the cooler-burning gas-turbine engines.

The rich-running spark-ignition engine operates at about 15 lbm of air per lbm of fuel, while in the leaner gas-turbine engine the air/fuel ratio is more like 40/1. Now, air is about 80% nitrogen, which does not enter into the reaction very much; so the combustion gases also contain mostly nitrogen. In view of these factors, it is often assumed in engine-cycle

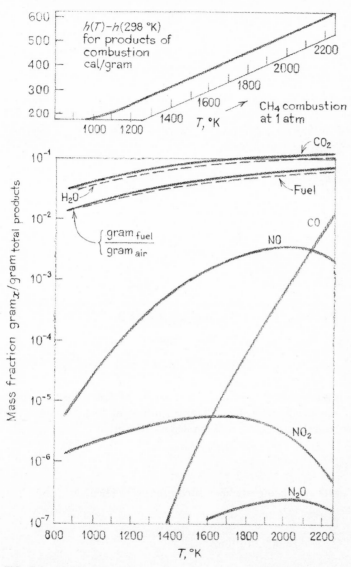

FIG. 5.6 PRODUCTS OF COMBUSTION FOR METHANE-AIR REACTION.

analysis that the working fluid is nitrogen (or air) and that the combustion process is simply a heat-transfer process that increases the energy of the working fluid. The amount of equivalent energy input as heat depends upon the chemical bonding energy released by the combustion process, which is very nearly the same as the heat of formation of the fuel. So,

FIG. 5.7   SIMPLIFIED COMBUSTOR MODEL.

as a first approximation, the amount of fuel required can be computed quite simply. For example, suppose we want to heat 10,000 lbm/hr of air from 500 to 2000°R in a steady-flow combustion chamber (see Fig. 5.7). The energy balance on the air is

$$\dot{Q} = \dot{M}_{air}(h_2 - h_1)_{air} \tag{5.10}$$

The enthalpy change of the air is related to the temperature change by Eq. (4.5),

$$h_2 - h_1 = c_p(T_2 - T_1)$$

So,

$$\dot{Q} = \dot{M}c_p(T_2 - T_1) = 10,000 \text{ lbm/hr} \times 0.24 \text{ Btu/lbm-°F}$$
$$\times (2,000 - 500)°F$$
$$= 3.6 \times 10^6 \text{ Btu/hr}$$

Now, high-grade hydrocarbon fuels have heats of formation (or combustion) in the neighborhood of 18,000 Btu/lbm of fuel. So, with the approximation outlined above, the equivalent heating is

$$\dot{Q} = \dot{M}_{fuel}Q_f$$

where we use Q_f to denote the heat of combustion. Then,

$$\dot{M}_{fuel} = \frac{\dot{Q}}{Q_f} = \frac{3.6 \times 10^6 \text{ Btu/hr}}{18,000 \text{ Btu/lbm}} = 200 \text{ lbm/hr}$$

Thus the required fuel flow is 145 lbm/hr, which corresponds to an air/fuel ratio of

$$AF = 10,000 \text{ lbm air}/200 \text{ lbm fuel} = 50 \text{ lbm air/lbm fuel}$$

This example shows the sort of simplified calculation that an engineer would make in order to establish a first estimate of the fuel requirements in a particular system. Table 5.1 gives some values for the heat of combustion for several fuels. These heats of combustion assume that the water formed by the reaction is gaseous, and are therefore called "lower heating values." The term "higher heating value" is used when the water is in liquid form, the small difference (a few percent) representing the amount of energy used in vaporizing the water in the products.

Table 5.1
Nominal Heats of Combustion of Selected Fuels

Fuel	Q_f, Btu/lbm
Hydrogen	52,000
Methane	24,000
Gasoline	21,000
Kerosene	20,000
Oil	18,000
Animal fats	17,000
Coal	12,000
Buckwheat hulls	7,500
Oak wood	7,000

Steady-flow combustion chambers, such as those used in boilers, gas-turbine engines, rocket motors, and household water heaters, come in a variety of designs. Burners using gaseous fuels usually mix the gas with some air to form a mixture that can be ignited easily. The mixture flows to a "flameholder," where it passes through the "flame front" and into the combustion zone of the flame. The products of combustion leave the flame region and are mixed with additional air, which reduces the temperature of the gases leaving the burner. Low-power burners, such as gas stoves, usually operate with a "laminar" flame, while in higher-power units the flame is purposely made "turbulent" to increase the mixing and assist the combustion process. Burners that use liquid fuels sometimes squirt the fuel directly into the combustion zone, and sometimes use heat drawn from the burner to vaporize the liquid before it enters the combustion zone; camp-stove burners are of this type. Solid-fuel burners generally draw the air along the burning fuel, and combustion occurs in a flame attached to the solid. Figure 5.8 shows a large commercial burner capable of handling either oil or gas, and Fig. 5.9 shows the combustion chamber from an aircraft gas-turbine engine.

In rockets the oxidizer is not air but instead is some chemical that can completely react with the fuel (no sense carting around nonreacting stuff!). Oxygen, stored in the rocket in a liquid form, is often used. Solid-propellant rockets have the fuel and oxidizer mixed together, and are

FIG. 5.8 INSTALLATION OF OIL/GAS BURNERS IN A LARGE UTILITY PLANT. COURTESY OF THE BABCOCK AND WILCOX CORPORATION.

shaped to produce a desirable pattern of burning and material removal from the motor.

In piston engines the combustion process is intermittent, and this causes special problems. For rapid and well-controlled ignition the mixture must be slightly rich, and this produces high flame temperatures, as mentioned previously. Ideally the flame initiates at the spark and propagates

Fuel injector   Flame holder

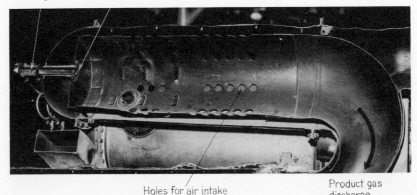

Holes for air intake

Product gas discharge

FIG. 5.9 CUTAWAY OF AN AIRCRAFT-ENGINE COMBUSTION CHAMBER.

across the cylinder at a steady rate. But if the combustible mixture is too hot (remember it gets hot as it is compressed), spontaneous ignition ("detonation") will occur at other places in the cylinder; this sends out uncontrolled flame fronts and shock waves, and produces undesirable local concentrations of high-pressure gases. This is the "pinging" or "knocking" that occurs in high-compression engines when a low-grade gasoline is used. The "octane" rating of a fuel is a measure of its ability to resist spontaneous ignition; high-octane fuels contain certain additives, and will make a high-compression engine run better. The energy released by high-octane fuel is essentially the same as that released by low-octane fuels; so it is a fabrication of the advertising agencies that "high-test" fuels have "more power." If your automobile engine runs without pinging on low-octane fuels, you will have absolutely no gain by using a high-octane fuel (except a larger gas-tax deduction for your income tax return).

An important advance in internal-combustion engine technology is the "staticharge engine" shown in Fig. 5.10. A conventional spark-ignition engine is modified by the addition of a "precombustion chamber" or "precup," which has its own supply of air and fuel and in which the sparking occurs. The idea is to introduce a rich, easily combustible mixture to the precup, ignite it, and then use the hot precup exhaust to ignite a much leaner mixture that has been admitted separately to the main cylinder. The overall effect is to permit the engine to operate much leaner than conventional internal-combustion engines, with air fuel ratios of the order of 25:1 and as high as 40:1. This virtually

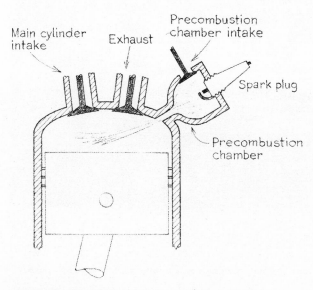

FIG. 5.10 THE STRATICHARGE ENGINE CONCEPT.

eliminates the problem with unburned hydrocarbons, drastically reduces the amount of carbon monoxide in the exhaust, and facilitates treatment of the oxides of nitrogen. The straticharge engine development is an interesting story with significant implications. The concept has been known for many years, and is used routinely in diesel engines. In the mid 1950s, Ralph Heintz and Professor A. L. London of Stanford University experimented with several straticharge engines of Heintz's design. Professor London wrote several papers proclaiming the advantages of straticharge engines as low-emission devices and pointing out the need for further straticharge engine development.* The concept was not pushed by U.S. automobile manufacturers, but Honda of Japan did carry out a massive development activity, which by 1973 resulted in an engine quite superior to anything that Detroit could offer. Licenses from Honda were taken by several U.S. manufacturers, and there was renewed interest in the straticharge engines under study at Stanford University (Fig. 5.11). In 1973 a study by a committee of the National Academy of Sciences again reached the conclusion that the diesel engine and the straticharge engine were the best bets for clean internal-combustion engines, *precisely* the conclusions given by Professor London in his

FIG. 5.11 AN EXPERIMENTAL STRATICHARGE ENGINE UNDERGOING TESTS IN THE MECHANICAL ENGINEERING LABORATORY AT STANFORD UNIVERSITY.

* Vehicle Smog Emissions, *Archives for Environmental Health*, vol. 6, May 1963, pp. 672–677.

pioneering paper ten years earlier, a paper that only Japanese industry really appreciated!

A number of new fuels are receiving attention today. A firm in California plans to produce electrical power by burning garbage, farmers in China use gas collected from farm-animal excrement for household cooking, and a British engineer runs his car on chicken droppings. Thermodynamic data for these fuels are not as well known as for conventional fuels, and the technology of their use needs to be developed; so there are some interesting and novel paths for combustion engineers to follow.

BATTERIES AND FUEL CELLS

A battery is a device for producing electrical energy from chemical energy. The "voltaic cell" (Fig. 5.12) consists of copper and zinc plates immersed in dilute sulfuric acid (the "electrolyte"). The electrolyte contains positive and negative ions, in this case H_2^{++} and SO_4^{--}, but overall is electrically neutral. Zinc tends to dissolve in the electrolyte more easily than does copper. In solution it forms the ion Zn^{++}, leaving electrons behind on the zinc plate. This gives the zinc plate a negative charge with respect to the copper plate, and so when an electric circuit is connected to the terminals of the cell, the electrons want to move around to the copper plate. Energy can then be extracted from the electrons in the external circuit. The battery becomes "dead" when the electrolyte is saturated with zinc, so that no more wants to dissolve. To revive the cell, one just replaces the sulfuric acid; this can be continued so long as zinc is present, but when the zinc is gone, the cell is permanently dead. The chemical reactions in the cell produce hydrogen around the copper plate,

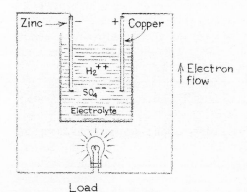

FIG. 5.12 THE VOLTAIC CELL.

and this must be removed for proper operation of the device. Another type of battery is the "dry-cell," in which the electrolyte is a paste instead of a liquid. The paste electrolyte and cell design are such that the electrolysis that produces hydrogen on the copper in the voltaic cell is not a problem.

The batteries described above are "primary" cells; they are stocked up with chemicals. The amount of chemical energy available is determined by the amounts and choices of initial chemicals, and the batteries are dead when the chemicals are used up. In contrast, the typical automotive battery and the nickel-cadmium dry-cell batteries used in transistor radios are rechargeable. The energy is put into the battery as electrical energy, which causes a chemical reaction to occur. The energy is then stored in the battery in chemical bonding energy, and is released upon demand by connecting up the cell. In the automotive battery the plates are lead and the electrolyte is dilute sulfuric acid. A layer of lead sulfate forms on these plates. When the battery is charged, chemical reactions occur that convert one set of plates to lead dioxide and the other to lead. The lead and lead dioxide plates then function as do the copper and zinc plates in the voltaic cell. The chemical energy stored in the plates is reconverted to electrical energy, which is extracted by an external load, and the plate surfaces are reconverted to lead sulfate.

The voltage output of a voltaic cell depends upon the chemicals used and not upon the size of the battery. The automotive-type lead-acid cell has a voltage per cell of about 2 volts, while typical carbon-zinc dry cells used in flashlights have a voltage of about 1.5 volts. Of course, cells can be connected in series to increase the total voltage if desired. The size of the cell determines the amount of electric current that can flow, i.e., the total amount of energy that is available from a single discharge of the cell. Strictly speaking, a "battery" is a collection of two or more cells, although the term is commonly used for single cells. The efficiency of modern batteries, defined as the ratio of the electrical energy output to the available chemical energy, is around 75%. The theoretical efficiency for *reversible* chemical reactions is, of course, 100%.

Let's think for a moment about the performance characteristics of a battery. Suppose we would like to have a battery pack that delivers a given amount of energy. The *energy* capabilities of the battery only depend only on its chemical composition. But what happens if we try to withdraw this energy rapidly? The *power* capabilities of the battery are limited by the ion-diffusion processes in the electrolyte; so, if we want to get the energy out quickly, we will need to have a lot of plate surface area, i.e., we will need to have a large battery of batteries. In other words, the higher the power output that we demand, the more pounds of batteries

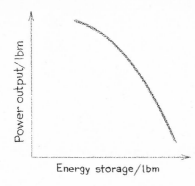

FIG. 5.13 TYPICAL BATTERY CHARACTERISTICS.

we will need to be able to get a given amount of energy. Thus, the energy delivered per lbm of battery will go down as the power output of each lbm is increased. This means that batteries must have performance curves of the form shown in Fig. 5.13; a battery can provide a small amount of energy at high power, or a larger amount of energy at low power.

There are many kinds of batteries, each with its own characteristics. They are compared on a specific-power–specific-energy basis in Fig. 5.14. The lead-acid battery is relatively well developed; it is rugged, efficient, and fairly economical. But its low specific energy makes it unsuited to

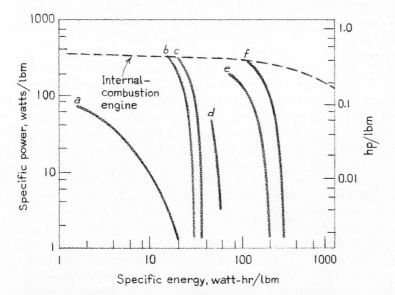

FIG. 5.14 BATTERY CHARACTERISTICS (1970). (*a*) LEAD-ACID, (*b*) NICKEL-CADMIUM, (*c*) NICKEL-ZINC, (*d*) ZINC-AIR, (*e*) SODIUM-SULPHUR, (*f*) LITHIUM-CHLORINE.

many applications, including electric automobiles and space vehicles. Nickel-cadmium batteries are now commonly used in battery-powered electric appliances. They are also rugged, and have a lifetime of a few thousand duty cycles as opposed to a few hundred for lead-acid batteries. A serious drawback is the relatively high cost of cadmium. Silver-zinc batteries have even greater specific energies but also suffer the disadvantage of high materials cost.

A number of new batteries are under development, and representatives of the high technology of batteries are included on the right-hand side of Fig. 5.14. A sodium-sulfur battery developed by the Ford Motor Company uses relatively inexpensive materials and has a specific energy an order of magnitude higher than lead-acid cells. But it must be operated at high temperatures (300°C), which causes operational difficulties. An attractive feature is that it can be charged at about the same rate that it can be discharged, which is unusual for rechargeable batteries. The lithium-chlorine battery under development at General Motors uses a molten lithium-chloride electrolyte at around 650°C, and its contents are rather hazardous. If the technical problems of safe containment can be solved, the high specific energy of the lithium-chlorine battery could be a factor that would spur its introduction to the consumer market, possibly as an energy source for electrical automobiles.

Other experimental batteries using organic electrolytes have been built; these seem to be limited to relatively low specific powers because of the low electrical conductivity of the electrolyte; but who can say what another dozen years of research on such devices might bring?

The "fuel cell" is another device in which chemical energy is converted directly to electrical energy. But unlike the battery, the fuel cell continuously takes in chemical fuels and discharges chemical wastes, and so is more like a steady-flow engine. The thermodynamic idea of the fuel cell is to capture the chemical bonding energy as fully available electrical energy *before* the energy has the chance to randomize into thermal energy. Thus, the fuel cell is *not* a heat engine, and is not limited by the Carnot efficiency; indeed, fuel cells theoretically can have efficiencies as high as 100%, and practical devices have efficiencies of the order of 60% (energy output/chemical energy input). A schematic of the hydrogen-oxygen fuel cell is shown in Fig. 5.15. Gaseous hydrogen and oxygen enter the cell at pressures of the order of 40 atmospheres and are brought into contact through porous electrodes. A liquid electrolyte between the electrodes serves to limit the reaction rate and hence to provide the desired control. Hydrogen diffuses through one electrode (called the "anode"), is absorbed on the surface, and then reacts with OH^- ions in the electrolyte

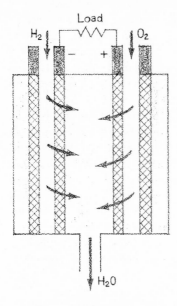

FIG. 5.15 FUEL CELL.

to form water and free electrons. Oxygen diffuses through the other electrode (called the "cathode") and reacts with the water to form OH^- ions. The electrolyte controls the migration rate of OH^- ions across the cell. The electrons flow out of the cell from the anode through the load, and energy is taken from them in the load before they are returned to the cathode. Water, the waste product of this fuel cell, must be continually removed. Imagine; an engine in which water is the waste product instead of carbon monoxide! Fuel cells of this sort have been used on many manned space missions, where the "waste" water can be put to good practical use.

There are many types of fuel cells under development today. The voltage of a cell is determined by the chemicals that are used as fuel, and most cells have potentials around 1 volt. The main reason why hydrogen-oxygen fuel cells are not in common use today is the fact that they require catalytic surfaces on the electrodes (a "catalyst" is a cupid-like material that does not really enter a chemical reaction but helps it to take place), and platinum (very expensive!) is the best catalyst. There have been numerous experiments with other types of fuel cells. One of the attractive features of the lithium-chlorine battery discussed earlier is that it can operate as a fuel cell. There is special interest in fuel cells that use air as the oxidizer. In particular, the hydrogen-air fuel cell seems very attractive. The hazard problem of storing hydrogen can be solved by using a "reformer." In this scheme a hydrocarbon fuel is allowed to react with hot steam in a catalytic reactor to produce hydrogen, which is then used in the fuel cell. The technology of large reformers is fairly well developed, but smaller units, such as might be required for automotive use, need further development. Fuel cells that use hydrocarbon fuels directly also have received some attention. Demonstration vehicles have been powered by hydrazine (a rocket fuel) and air cells, which can start up from room temperature and operate at under 100°C. If certain technological problems with the catalyst can be solved, and if the price of hydrazine can be reduced by 95%, this system could become a real contender in the electric-automobile market. Other fuel cells using methanol and air, or liquid petroleum gas (LPG) and air, are now under development. The advantages of being able to use low-grade hydrocarbon fuels directly in a room-temperature fuel cell are obvious, and the concept is so attractive that there is great interest in long-term research aimed at its development.

Let's think about the performance characteristics of fuel-cell systems. Suppose we want to obtain a certain amount of energy from the system, as, for example, required by a certain period of automobile operation. The more energy we desire the more fuel we will have to put on board, and hence the greater will be the system weight. However, if we ask for just a little energy, then the main contribution to the system weight will come from the hardware and not the fuel. Now, a given fuel-cell system will have a given power capability. Considering all these factors, we reason that for *small* energy demands the specific power (power per system lbm) should be independent of the total energy to be delivered from one fueling. However, if we desire to increase the range of the vehicle, we will have to take on more fuel, and thus the specific power will drop as the specific energy (energy per system lbm) increases. When the fuel has increased to the point where its weight controls the system weight, the specific energy will be virtually constant, determined only by the fuels used and the efficiency of the full cell. This reasoning explains the form of the fuel-cell performance curves shown in Fig. 5.16.

Table 5.2 shows some projections of fuel-cell technology as it might exist around 1980. Although at this writing fuel cells have not yet developed to the point where they can move into consumer markets in any major way, it does seem likely that their continued development will make this a distinct possibility within a decade or two. As energy technology moves on in the years ahead, fuel cells will probably become everyday sources of clean energy for automobiles, homes, and perhaps even cities.

Table 5.2
The Outlook for Fuel-cell Technology*—1980

Fuel	Temp., °C	Voltage per cell	Cell efficiency	System efficiency	System cost, $/kw	Life, hr	Fuel cost, $/kwhr
Hydrogen/air	80	0.7	0.47	0.43	100	1,000	1.43
Hydrogen/air with LPG reformer	80	0.7	0.47	0.35	150	1,000	0.46
Hydrogen/air with kerosene reformer	80	0.7	0.47	0.35	180	1,000	1.21
Hydrogen/air with ammonia reformer	80	0.7	0.47	0.38	120	1,000	0.40
Hydrazine/air	80	0.7	0.44	0.40	100	1,000	143
Methanol/air	80	0.5	0.40	0.33	3,000	1,000	0.47
LPG/air	200	3.0	0.42	0.38	1,000	1,000	0.42

* "The Automobile and Air Pollution," Part II, U.S. Department of Commerce, December 1967.

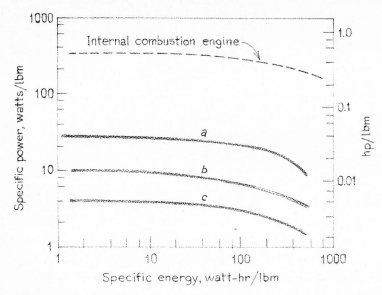

FIG. 5.16 FUEL-CELL CHARACTERISTICS (1970). (*a*) HYDROGEN-AIR, (*b*) KERO-SENE-AIR, (*c*) METHANE-AIR.

CHEMICAL ENERGY: STORAGE, TRANSPORTATION, AND TRADE

Chemical energy is a form that is particularly well suited to both shipment and storage; the bonding energy available in 1 lbm of gasoline is about 18,000 Btu. In contrast, the potential energy of 1 lbm of water at an elevation of 100 ft is only about 0.13 Btu! The vast oil and natural gas shipments that take place worldwide attest to the importance of chemical energy in energy transportation. The other significant energy transportation mode is electrical, but there are no known ways to store large amounts of electrical energy cheaply.

There have been proposals for replacing our electrical-energy transportation system by a chemical-energy transportation system using hydrogen or alcohol as the "energy currency." Hydrogen has the advantage of high-energy density (52,000 Btu/lbm), can be made by electrolysis of water, and can be used in fuel cells to produce electrical energy. Alcohol can be produced from agricultural fuels, and thus looms large in proposals for use of agriculture for solar energy conversion.

These considerations make it clear that chemical energy will remain very important in the years ahead; new chemical-energy commodities will certainly become important in international trade. Countries that can learn to produce chemical fuels cheaply using solar or nuclear energy may well dominate the international scene in the next century.

PROBLEMS

Balancing the atoms

5.1 Complete the chemical equation for the combustion of *ethane*.

$$C_2H_6 + \_\_\_O_2 \rightarrow \_\_\_CO_2 + \_\_\_H_2O$$

5.2 Complete the chemical equation for the combustion of *hexylene*,

$$C_6H_{12} + \_\_\_O_2 \rightarrow \_\_\_CO_2 + \_\_\_H_2O$$

To develop your skill with numbers

5.3 The atomic weights of carbon, hydrogen, and oxygen atoms are

C   12 grams/gmole
H   1 gram/gmole
O   16 grams/gmole

Compute the atomic weights of *ethane* (C_2H_6) and *hexylene* (C_6H_{12}).

5.4 Do Probs. 5.1 to 5.3. Then, compute the number of grams of oxygen per gram of fuel for the two reactions.

5.5 For each mole of O_2 in air there are 3.76 moles of N_2. The resulting molecular weight of air is about 29 grams/gmole. Do Prob. 5.4, and then compute the air/fuel ratio for the indicated reactions.

5.6 The heat of combustion of ethane is 386 kcal/gmole and for hexylene 953 kcal/gmole. Calculate the heat of combustion in Btu/lbm for each fuel (see Prob. 5.3).

5.7 Do Probs. 5.1 to 5.6. Then assuming that the products of combustion have an average c_p of 7 cal/gmole-°K, calculate the adiabatic flame temperature, assuming that the reactants are at room temperature, for the two reactions discussed.

5.8 A power plant with an overall efficiency of 30% produces 1 Gw of electrical power using coal ($Q_f = 12,500$ Btu/lbm) as the fuel. Calculate the annual requirements for coal in tons.

5.9 A gas-turbine engine produces 5,000 hp for industrial use. The engine has an efficiency of 25% and uses kerosene ($Q_f = 20,000$ Btu/lbm) as the fuel. Calculate the hourly fuel consumption in lbm.

5.10 A farmer wants to generate electricity by burning trash. He can purchase a small steam-boiler turbine and generator and the plant will have an overall efficiency of 12%. He wants to generate a total of 30,000 kwhr each year. Calculate the tonnage of trash required, assuming that $Q_f = 7,000$ Btu/lbm.

Some more interesting analyses

5.11 It has been suggested that the United States shift back to use of wood as the major fossil fuel. A massive forest-station project is proposed, in which trees would be replanted in a 50-year cycle, and cut for fuel. The forests would provide additional recreational facilities prior to harvesting, and the planting program would employ thousands of low-skilled workers. Estimate the annual demand for wood fuel in lbm, assuming that 2×10^{12} kwhr are generated in steam power plants with 35% efficiency. Estimate the *volume* of wood involved (wood has a density of around 50 lbm/ft³) in "board-feet" ("board-foot" is a volume 1 ft square by 1 in thick).

Look up the board-footage yield per acre of timber in the northwestern United States, and calculate the acreage that would have to be harvested annually to supply the electrical demands. What total acreage would have to be under cultivation for this system? Discuss the feasibility of this plan, technical difficulties that might be involved, the impact of the plan on other elements of the economy (e.g., transportation), and the social and political values that would be relevant in a public discussion of the proposal.

5.12 It has been proposed that garbage be burned for production of electrical power. Pick a large city in the United States; find out its annual production of garbage in tons, and its annual consumption of electricity in kwhr. Estimate the electrical energy that could be produced from the garbage, and compare that with the electrical needs. Discuss the social, economic, and political questions that would be important in discussion of, implementation of, or defeat of this plan. Based on your values and your quantitative analysis, what do *you* think of the plan?

5.13 One problem with solar energy is that it is not available at night. One possible alternative is to use solar energy to generate electricity during the day, to use this electricity to decompose water into hydrogen and oxygen, and then to combust the hydrogen with the oxygen over the 24-hr period to provide heat for the continual production of electricity.

The system will consist of a steam power plant that burns hydrogen to generate electricity, and another that uses the solar energy to generate electricity during the day to generate just enough electrical energy to decompose the hydrogen. The electrolysis process will require electrical energy of around 70 kcal/gmole, and the atomic weight of water is 18 grams/gmole. The solar energy plant will convert about 30% of the collected solar energy into electrical energy for the electrolysis process. The combustion of hydrogen will yield about 52,000 Btu/lbm, and the main power plant will convert about 35% of this energy into electrical energy for consumer use. About 100 Btu will be collected on each square foot of solar collector surface each hour during 10 daylight hours. Atomic-weight data are given in Prob. 5.3.

Calculate the following quantities for a system that delivers 1 Gw of electrical power all the time:

(*a*) Daily solar energy collection requirements, Btu
(*b*) Solar collector area
(*c*) Daily mass of water electrolyzed, lbm
(*d*) Daily hydrogen consumption rate, lbm

Assuming that the products of combustion are water, which is cooled in a dry cooling tower and returned to be electrolyzed again, calculate the volume of water that must be handled each day (ft^3). Discuss the technical problems that you foresee with this system, and the social, political, and economic questions that would arise in serious consideration of the plan.

5.14 A new coal reserve estimated to contain 10^{10} ft^3 of coal-bearing soil with an average coal density of 90 lbm/ft^3 has been discovered adjacent to a popular recreational area. Estimates suggest that the average heat of formation of the coal is around 13,000 Btu/lbm, and the utility companies in the area are seeking approval for initiation of open-pit mining activities. They claim that the new discovery will provide enough coal to run the nation's fossil-fuel power stations for 25 years or more. Discuss the situations quantitatively. Is this reserve worth developing at the possible expense of the recreational area? Is the recreational area worth saving in view of the vast amounts of energy that the coal reserve offers?

5.15 A proposed state law will outlaw power plant burners that operate with flame temperature less than 3,000°F. The idea is to force higher-efficiency power plants, in the interest of reducing the amount of pollutants from fossil fuels. The utilities companies claim that the law might actually increase the pollution problem. Sponsors of the law argue that the utilities companies are just trying to save their dying empires. As a politician seeking sound technical advice on the matter, what sort of consultant would you seek for technical advice, and what sort of analysis might you ask him to provide? How would you go about deciding which point of view is correct?

6

HOW ELECTRONS WORK

In which we see some
electrifying things

SOME BASIC ELECTRIC CONCEPTS

Newton's laws of motion and gravitation provided a scientific explanation for some but not all the forces observed in nature. They do not, for example, explain why a piece of amber when rubbed will attract pieces of straw, a phenomenon known to Thales of Miletus in 600 B.C., or why magnetite rocks will attract iron. Some two hundred years after Newton, Hans Oersted observed a connection between these two phenomena, and in the early 1800s a new science, "electromagnetism," was born. It took many years to get the science straight, and in fact the job is still not quite complete. Theoretical physicists are still trying to pull electromagnetics, gravitation, and quantum physics together into one grand physical theory;

but the Einstein to do it has not yet spoken. Inventors began to make useful electromagnetic devices before even the most basic concepts and principles of electromagnetics were understood. Indeed, the desire to understand the physics in order to make better devices was a prime motivating factor in the development of electromagnetic science. In this chapter we will acquaint you with those aspects of the science which will be most helpful to you in understanding energy systems.

In order to explain the various phenomena of electromagnetics, scientists introduced several new basic concepts. The most basic concept in electromagnetics is "electric charge." Charge is conceived as a *conserved property* of electric particles; charge cannot be created or destroyed. It seems to come in very small but discrete amounts of two different types called "positive" and "negative." Forces exist between charges; charges of the same type repel one another, and charges of opposite type attract one another. The magnitude of the force between two charges decreases as the charges are separated. Part of the concept of electric charge is that the total electric charge of an object is the sum of the charges of all particles in or on the object. An object with zero total charge is called electrically "neutral."

To be really useful, the concept of charge must be made quantitative so that values of charge can be measured and magnitudes of electric forces can be predicted. Careful experiments show that the force between fixed charges varies inversely as the square of the distance between them, and this leads to "Coulomb's law,"

$$F = k_c \frac{Q_1 Q_2}{r^2} \tag{6.1}$$

Here Q_1 and Q_2 are the values of the two charges, r is the distance between them, k_c is a constant, and F is the force (positive for repulsion, negative for attraction) between the charges. Figure 6.1 shows this law schematically. Just as Newton's law provides the fundamental way to think about mass, Coulomb's law provides the fundamental way to think about charge. The value and units of k_c depend upon the units of measurement

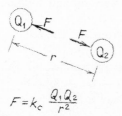

$$F = k_c \frac{Q_1 Q_2}{r^2}$$

FIG. 6.1 COULOMB'S LAW.

for charge. It is possible to set up unit systems in which $k_c = 1$, analogous with the treatment of Newton's law in the mks system [see Eq. (2.7)]. In such a system charge has the dimensions of $\sqrt{\text{force}}$/distance. Alternatively, one can set up a measurement standard for charge independently, and then k_c will not be unity; this is analogous with the "engineering" treatment of Newton's law [Eq. (2.9)] and is the more modern approach. The current international system uses the "coulomb" as the unit of charge and the mks system for forces and distances. You can think of the coulomb as being defined in terms of the charge on an electron, with the electronic charge having the value of -1.6021×10^{-19} coulombs. Then, k_c has the experimentally determined value of 8.99×10^9 newton-meter^2/coulomb^2.

To illustrate Coulomb's law, let's compute the force between two electrons one "micron" (10^{-6} meters) apart:

$$F = 8.99 \times 10^9 \frac{\text{newton-m}^2}{\text{coulomb}^2}$$

$$\times \frac{(-1.6 \times 10^{-19} \text{ coulomb})(-1.6 \times 10^{-19} \text{ coulomb})}{(10^{-6} \text{ m})^2}$$

$$= 2.3 \times 10^{-16} \text{ newton}$$

or about the weight of one cell in that hamburger that weighs a newton.

Another example will help establish a quantitative feel for the coulomb as a quantity of charge. Let's compute the charge required to develop a force of 1 newton between two equal charges separated by 1 meter. Using Eq. (6.1),

$$Q = r\sqrt{\frac{F}{k_c}} = 1 \text{ m } \sqrt{1 \text{ newton}/(8.99 \times 10^9 \text{ newton-m}^2/\text{coulomb}^2)}$$

$$= 1 \times 10^{-5} \text{ coulombs}$$

You can calculate for yourself that a charge of -10^{-5} coulombs is equivalent to the charge of 6.2×10^{13} electrons. So, lots and lots of electrons a few feet apart will give rise to only a little shove.

The previous calculation assumes that all the electrons are at the same point. But, since like charges repel, the electrons will tend to push apart from one another and distribute themselves around the surface of any body on which they are placed. If a "test" charge is moved around near the body, the charge will experience forces that depend in a much more complicated way on its position. One could compute the force on the test charge by summing the force contributed by each electron on the body, but with 10^{13} electrons to consider this is a lot of summing. It is more convenient to use another concept called the "electric field." Since the force on the test charge depends only upon the magnitude of the test charge and its position, one can write

$$F(x) = QE(x) \tag{6.2}$$

where Q is the magnitude of the test charge and $F(x)$ is the force; the notation $F(x)$ shows that F depends on the position of the test charge, denoted by x. The factor $E(x)$ is the "electric field," which also depends upon position as indicated by the x. You can see that E has the dimensions of force/charge, and hence the units of newtons/coulomb. The electric field represents the total push of all other charges on a unit test charge. Since the push F has a direction (in the language of mathematics it is a "vector"), the electric field E also has a direction (it too is a vector), namely, the direction of the push on Q. Figure 6.2 displays this idea. According to Newton's law, a free body will accelerate in the direction of an applied force. So, charged particles will accelerate in the direction of the electric field; this has important application in a variety of practical systems, including xerox copiers, dust precipitators for air-pollution control, and television tubes.

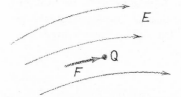

FIG. 6.2 THE ELECTRIC FIELD EXERTS FORCES ON CHARGES.

The push that an electric field gives a charge will do work on the charge if the charge moves. If the push of the electric field is constant during this motion, the work done on the charge by the electric field will be

$$W = F \cdot L = QEL \tag{6.3}$$

where L is the distance that the charge Q moves in the direction of the electric field. If you think about it, the path from one point to the other could be anything, and Eq. (6.3) would still hold, because there are no forces on the particle perpendicular to the electric field; Fig. 6.3 shows

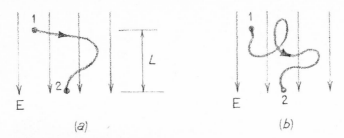

FIG 6.3 THE WORK DONE
DEPENDS ONLY ON THE POSITIONS
1 AND 2, AND NOT ON THE
CHOICE OF PATH.

this idea. This means that the work done does not depend on the path taken from point 1 to point 2, but instead depends only upon the location of points 1 and 2. The product EL is called the "electric potential" or "voltage" difference between points 1 and 2. One can pick an arbitrary reference point, and then define the voltage (relative to this reference) for all other points. Denoting the voltage by V, Eq. (6.3) can be expressed as

$$W = Q(V_2 - V_1) \qquad (6.4)$$

The work done by the electric field on the charge is therefore simply the value of the charge times the difference in voltage between the stopping and starting points. The voltage V has the dimensions of energy/charge, and hence the units of newton-meters/coulomb. This complicated set of units is given the nickname "volts,"

1 volt $=$ 1 newton-meter/coulomb

You may have had some previous experience with a voltmeter, which reads the magnitude of the electric potential difference between two points.

Let's now consider a familiar 1.5-volt dry-cell battery, in which positive charges (ions) collect on the positive terminal and electrons on the negative terminal. The electrons don't move very easily through air, but if an electrical "conductor," such as a copper wire, is used to connect the positive and negative terminals, electrons will move freely through the wire. An electron in the wire will experience a force attracting it toward the positive terminal and repelling it from the negative terminal. The electric field in the wire will be in the direction shown in Fig. 6.4. Electrons, being negatively charged particles, will tend to move in the opposite

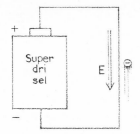

FIG. 6.4 ELECTRIC FIELD AND
ELECTRON FLOW FOR A DRY-CELL
BATTERY.

direction of the electric field, i.e., from low voltage to high. So, if the electron manages to move through the wire from one terminal to the other, the work done by the electric field on the electron will be

$$W = Q(V_2 - V_1) = (-1.6 \times 10^{-19} \text{ coulomb})(-1.5 \text{ volts})$$
$$= 2.4 \times 10^{-19} \text{ joules}$$

Note that $V_2 - V_1$ is negative because the electron moves opposite to the direction of the electric field. The unit "joule" is shorthand for newton-meter, which you get when you multiply coulombs times volts if you remember that volts stands for newton-meter/coulomb. What happens to this energy? In a copper wire the electron will simply bounce around as it fights its way through the array of the wire's atoms to reach the positive terminal. These collisions transfer energy from the electrons to the copper atoms in a random way (entropy is produced), and the wire gets hot. The electric field is produced by the positive and negative charges in the battery. So, energy flows out of the battery whenever its electric field does work.

Another way to look at Eq. (6.4) is to say that the voltage is the "potential energy per unit of charge." Then, the work done by an electric field as it pushes charges is equal to the increase in potential energy of the charges.

STEADY ELECTRIC CIRCUITS

Let's consider an entire stream of electrons moving along in a wire that forms part of a closed circuit (Fig. 6.5). The "electric current" is defined as the *positive* charge flow per unit of time across a section of the wire. Recalling that the electrons have *negative* charge, the electron motion to the left means in effect that the current is flowing to the right. If Q is the

FIG. 6.5 ELECTRIC-CURRENT FLOW.

total (positive) charge that flows across the section and t is the time required for this to happen, the current I is

$$I = \frac{Q}{t} \tag{6.5}$$

Current therefore has the dimensions of charge/time and the units of coulombs/sec. For short this is called "amperes,"

1 ampere = 1 coulomb/sec

An ammeter measures the rate of charge flow in "amps," which is short for amperes. You are probably familiar with fuses. A "fuse" is a device which gets so hot that it melts when the current flow through it reaches a certain amperage. When it melts, the closed electric circuit is opened, and the current ceases to flow.

Suppose we have a steady current flow in a wire, and a reference position in the electric field that we have defined as "zero volts" (usually this is taken as the earth, and hence zero volts is called "ground"). If the current should pass through a device and return to ground, the work done by the electric field on the charges forming the current would be

$$W = QV = ItV$$

where V is the voltage difference across the device and again I is the current flow. This is energy that could be converted to useful work in the device. Thus, we can regard the product QV as the *energy* of the charge. Then, the *power P* of the current flow is the energy flow per unit time, or

$$P = IV \tag{6.6}$$

Thus for a steady current ("direct current" or "dc") the product of the potential (volts) and the current (amps) gives the *power* flowing with the current. The units of power are volt-amps, which is called "watts," which is shorthand for joules/sec, which is shorthand for newton-meter/sec, which is shorthand for . . . well, you can figure it out in terms of meters, kilograms, and seconds.

For example, suppose we have a pair of large wires that carry steady current to and from a "load" (the term commonly used for an expender of electrical energy). Suppose one wire has a potential of $V_1 = 10,000$ volts and the other wire is "neutral" or "grounded" ($V_2 = 0$ volts). If the circuit carries a current of 2,000 amps, the power being delivered to the load in the high-voltage wire is

$$P = 10{,}000 \text{ volts} \times 2{,}000 \text{ amps} = 2 \times 10^7 \text{ watts} = 20 \text{ Mw}$$

The power being removed in the low-voltage wire is zero, and thus by an energy balance 20 Mw is the net electrical power input to the load. This is a lot of power, and we hope that the load uses it constructively. Let's assume steady-flow steady-state conditions exist in the load, and use $\dot{W}$ and $\dot{Q}$ to represent the power outputs from the load as work and heat, respectively. Since charge is conserved, the current flows in and out of the load will be identical. Then, an energy balance on the load (Fig. 6.6) gives

$$\underset{\substack{\text{energy} \\ \text{input} \\ \text{rate}}}{IV_1} = \underset{\substack{\text{energy} \\ \text{output} \\ \text{rate}}}{IV_2 + \dot{W} + \dot{Q}}$$

Denoting the voltage difference $V_1 - V_2$ simply by V, we can interpret this energy balance as

$$\underset{\substack{\text{electrical} \\ \text{power} \\ \text{input}}}{IV} = \underset{\substack{\text{power} \\ \text{output} \\ \text{as work}}}{\dot{W}} + \underset{\substack{\text{power} \\ \text{output} \\ \text{as heat}}}{\dot{Q}} \qquad (6.7)$$

In this model the load converts part of the electrical power to useful power $\dot{W}$ and the rest to thermal energy ($\dot{Q}$). The "conversion efficiency" of the device could be defined as the ratio of the useful power output to the electrical power input. If the load is a complex of direct-current electric motors, the efficiency might be about 90%.

As current flows through a light bulb or the heating coils in a toaster, the electrons collide with the molecules in the conductor, raising the conductor temperature. Such loads are called "purely resistive"; they do not convert electrical energy to useful work but instead convert it all

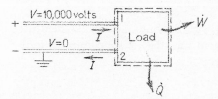

FIG. 6.6 ENERGY BALANCE ON AN ELECTRICAL LOAD.

to thermal energy. Since for many loads the ratio of the voltage difference to the current flow is independent of the voltage, it is useful to define the "resistance" R of a load by

$$R = \frac{V}{I} \qquad (6.8)$$

Here V is the voltage difference across the load and I is the current flow through it. Equation (6.8) is called "Ohm's law." If the load has a high resistance (large R), there will be a large voltage drop for a given current; if the load resistance is very low, a lot of current will flow for a given imposed voltage. The units of R are volts/amp, which is the same as newton-meters-seconds/coulombs², which is pretty involved; so the combination volts/amp is given the nickname "ohms," and resistances of loads are stated in ohms.

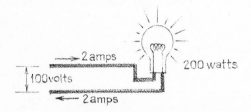

FIG. 6.7 FIGURING THE BULB CURRENT.

As an example, let's consider a light bulb that used 200 watts of electrical energy when connected to a 100-volt dc supply (Fig. 6.7). The steady current flowing through the bulb is given by Eq. (6.6) as

$$I = \frac{P}{V} = \frac{200 \text{ watts}}{100 \text{ volts}} = 2 \text{ amps}$$

From Eq. (6.8), the resistance of the bulb must be

$$R = \frac{V}{I} = \frac{100 \text{ volts}}{2 \text{ amps}} = 50 \text{ ohms}$$

Suppose there was something wrong with the bulb, and its filament wire had a resistance of only 10 ohms. From Eq. (6.8), if we connect it to the 100-volt supply, it will draw a current of

$$I = \frac{V}{R} = \frac{100 \text{ volts}}{10 \text{ ohms}} = 10 \text{ amps}$$

The power consumed by this bad bulb would then be

$$P = VI = 10 \text{ amps} \times 100 \text{ volts} = 1,000 \text{ watts}$$

In order to operate at steady state, the bulb would have to get sufficiently hot that it could get rid of all this input electrical energy as heat by radiation, conduction, and convection. Chances are that the temperature would go up and up and that the filament would melt before a steady-state operating temperature had been achieved.

One can combine Eq. (6.6) and (6.8) to obtain a third equation that relates the power P, the current I, and the load resistance R,

$$P = I^2 R \tag{6.9}$$

This shows that, for a given load resistance, the power that must be removed as heat varies with the *square* of the current. This power is, for obvious reasons, called the "I^2R" (pronounced "eye-squared-are") loss of the load. It occurs whenever a current flows through a resistance.

Let's now consider what happens in a transmission line. The power being transmitted is

$$P_t = VI$$

where V is the voltage difference between the "hot" and "ground" transmission lines and I is the current flow in the circuit. Solving for I,

$$I = \frac{P_t}{V} \tag{6.10}$$

Now, the transmission line will have some resistance R, and hence there will be some I^2R losses, or "dissipation," in accordance with Eq. (6.9). Thus, the dissipated power P_d is

$$P_d = \left(\frac{P_t}{V}\right)^2 R \tag{6.11}$$

This shows that, for transmission of a given amount of power P_t through a line of given resistance R, the transmission loss P_d goes down as the *square* of the line voltage V. Very high voltages, up to 500,000 volts, are used for electrical energy transmission lines to keep the I^2R losses small.

For a wire the resistance R is proportional to the wire length L and inversely proportional to the cross-sectional area A. The proportionality factor depends upon the wire material. The "electrical conductivity" of the material is usually denoted by the greek symbol κ ("kappa"), and is defined by the equation

$$R = \frac{1}{\kappa}\frac{L}{A} \tag{6.12}$$

The electrical conductivity bears the same relationship to electric current flow and voltage differences as the thermal conductivity bears to heat flow and temperature differences [see Eq. (3.7)]. The reciprocal of κ is termed the "resistivity" of the material, and is denoted by the Greek symbol ρ ("rho"). Thus, the resistance can be expressed as

$$R = \frac{\rho L}{A} \tag{6.13}$$

The resistivity has the dimensions of resistance times length, and is usually expressed in ohm-centimeters. Values of ρ for several materials are given in Table 6.1. Note that silver has the lowest resistivity; it is expensive. The next lowest is copper, which explains why it is used in electric circuits and wires. Porcelain has a high resistivity, which explains why it is used as an electrical insulator. There is also a big difference between mercury and whiskey, which explains why mercury and not whiskey is used in quiet switches and household thermostats.

An electric circuit is composed of good conductors, such as copper or other metals. Electrons will flow with relative ease through any conductor. However, if the circuit is "broken" by insertion of an insulator (including

Table 6.1
Resistivity of Selected Materials

Substance	Resistivity, ohm-cm
Silver	1.6×10^{-6}
Copper	1.7×10^{-6}
Aluminum	2.8×10^{-6}
Mercury*	1×10^{-4}
Whiskey*	2×10^{5}
Wood	10^{12}
Porcelain	10^{15}

* Liquid.

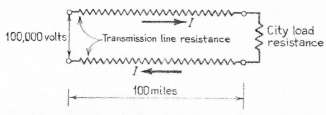

FIG. 6.8 COMPUTING LINE LOSSES.

air) through which electrons have enormous difficulty moving, so that the electrons cannot easily return from whence they came, they will bog down, and the current flow will effectively cease. So, there will be no steady-state current in an "open" electric circuit; for a current to flow, the electrons must have a closed conductor circuit all the way back to their source.

As an example, let's compute the resistance of a 100-mile length of copper wire 1 cm in diameter (Fig. 6.8). The cross-sectional area is

$$A = \frac{\pi d^2}{4} = 0.78 \text{ cm}^2$$

and the length is

$$L = 100 \text{ miles} \times 5{,}280 \text{ ft/mile} \times 12 \text{ in/ft} \times 2.54 \text{ cm/in}$$
$$= 1.6 \times 10^7 \text{ cm}$$

so the resistance is

$$R = \frac{\rho L}{A} = 1.7 \times 10^{-6} \text{ ohm-cm} \times 1.6 \times 10^7 \text{ cm}/0.78 \text{ cm}^2$$
$$= 35 \text{ ohms}$$

which is about the same as the resistance of a 40-watt light bulb. Suppose a pair of such wires are used in the circuit supplying electrical energy to a city at a potential of 100,000 volts. If the transmitted power P_t is 20 Mw = 2×10^7 watts, then the current is, from Eq. (6.10),

$$I = \frac{P_t}{V} = 2 \times 10^7 \text{ watts}/10^5 \text{ volts}$$
$$= 200 \text{ amps}$$

The I^2R line loss in each wire is

$$P_d = I^2R = (200 \text{ amps})^2 \times 35 \text{ ohms} = 1.4 \times 10^6 \text{ watts} = 1.4 \text{ Mw}$$

(remember that ohm means volts/amp and watt means volts times amps). Now the current must come in through one wire and go back to the powerhouse through the other; so the total line loss will be twice this amount, or 2.8 Mw. This is about 14% of the transmitted power! In order to reduce line losses, an engineer could specify larger wires or higher voltages. With larger wires the copper costs go up, the towers must be stronger to support the extra weight and taller for safety, and installation and maintenance of the heavier cable cost more. With higher voltages there can be arcing across insulators, electrical interference to radio receivers nearby, and other practical problems. As in any kind of engineering a compromise is required, and this compromise should be dictated by a mix of the technical and economic realities and the pertinent social values and priorities.

The line loss results in a drop in the voltage supplied at the load. Applying Ohm's law (Eq. 6.8) to one of the transmission lines itself as a load, voltage drop along each line will be

$$V = IR = 200 \text{ amps} \times 35 \text{ ohms} = 7,000 \text{ volts}$$

So, the net voltage difference supplied at the load will be

$$100,000 \text{ volts} - 2 \times 7,000 \text{ volts} = 86,000 \text{ volts}$$

When more than one load is to be connected to a single power supply, there are two basic approaches. If the loads are connected in "series" (Fig. 6.9a), the same current will flow through every load, the voltage across each load will be given by Ohm's law [Eq. (6.8)], and the sum of the voltage drops across each load will equal the supply voltage. If one of the loads burns out, the circuit is opened, and so the current stops. This is not very convenient; imagine having all your lights go out just because one bulb burned out! Moreover, not all loads can handle the same current. So, the series connection is never used in a power system. Instead a "parallel" circuit is used (Fig. 6.9b). Here each load is supplied with the same voltage, and the current that each draws is determined by its resistance in accordance with Ohm's law. The main wires near the power supply have to be able to carry the sum of the currents drawn by all loads, and so these leads will want to be heavier wires. You might compare the large wires bringing power into a home with the smaller

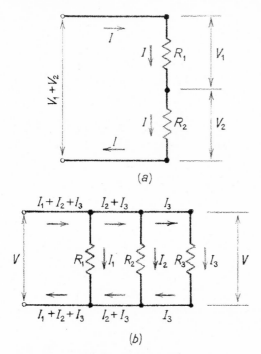

FIG. 6.9 (a) SERIES CONNECTION—
EACH LOAD CARRIES THE SAME
CURRENT. (b) IN A PARALLEL
LOAD CONNECTION EACH LOAD
SEES THE SAME APPLIED VOLTAGE.

wires used to distribute the power in several parallel circuits connected
to the supply at the fuse or circuit-breaker box.

People in the city would certainly not want household wiring at the
potential of 86,000 volts. How can the voltage be reduced to a level that
is safe for local consumption? One way is to use additional resistance in
series with the load. But this is wasteful of energy, for now in addition
to the I^2R losses of the load we have the I^2R of the control resistor.
In order to change the voltage of electrical energy from one level to another efficiently, the *unsteady* behavior of electricity must be exploited
through use of "alternating current ("ac"). We'll look at this in the next
section.

ALTERNATING ELECTRIC CIRCUITS

The notions of charge, Coulomb's law, and the electric field do not
provide complete explanations for all electric phenomena. We expect
that you are familiar with the notion of a magnetic field through previous

study or play with magnets. Magnets can pull or push on other magnets, and will pull on "magnetic materials" (defined as those materials which are attracted by magnets). Scientists use the concept of a "magnetic field" to explain these phenomena. A magnetic field is, by conception, sort of a force field, something like an electric field, except that a magnetic field by definition can exert forces only on *moving* charges. Since inventing the two ideas of electric and magnetic fields, scientists have been able to explain quantitatively and analyze all electromagnetic phenomena. This ability provides a strong experimental confirmation of the theory of electromagnetism.

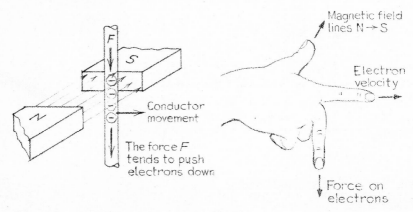

FIG. 6.10 MAGNETIC FIELDS EXERT FORCES ON MOVING ELECTRONS.

Figure 6.10 illustrates the effect of a magnetic field on a moving charge. The lines drawn connecting the north and south poles of the magnet represent the magnetic field. We show an electrical conductor moving through the space between the pole faces. The magnetic field will exert a force on the *moving* electrons in the conductor *perpendicular to both the magnetic field lines and the direction of motion*. This will push the electrons in the direction shown. The direction of the force exerted by the magnetic field on electrons can be conveniently remembered by the "right-hand rule" (see Fig. 6.10). The magnitude of the force depends upon the electron velocity, the strength of the magnetic field, and the angle between the direction of the magnetic field and the direction of the electron motion. If the electrons are free to move along the conductor, a current will flow, and then the magnetic field will be doing work on the electrons. This is the basic idea behind electric generators.

Magnetic fields themselves are produced by moving charges. The formation of a magnetic field by a current flow is shown in Fig. 6.11. A

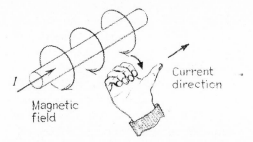

I

Magnetic
field

Current
direction

FIG. 6.11 CURRENTS PRODUCE MAGNETIC FIELDS.

current produces a magnetic field *perpendicular* to the direction of current flow in a manner determined by another famous "right-hand rule." Stick your right thumb in the direction of the current I, and your fingers will curl in the direction that the magnetic field points (see Fig. 6.10). If the wire is coiled, each winding will produce a magnetic field in accordance with the right-hand rule, and these will add up to give the total magnetic field. Various complex magnetic field patterns can be produced by winding the coils in various clever ways. This same idea is used to explain permanent magnets; one thinks of the electron circulation in the molecules as little current loops that produce the magnetism. Magnetic materials are those in which the moving electrons are attracted by an external magnetic field.

To summarize, stationary charges form an electric field that exerts forces on other charges; and moving charges form magnetic fields that exert forces on other moving charges. But suppose you are a scientist observing forces on charges from a vantage point on the *moving* conductor in Fig. 6.10. To you the charges in the wire are stationary (before F gets them moving), yet you see them as experiencing the force F. You can't regard F as arising from a magnetic field, because you and your fellow scientists have agreed that magnetic fields don't exert forces on stationary charges. So, you must attribute the force F to an *electric* field. If the wire is motionless between the poles, F will be zero. The faster the wire carries you through the space between the poles, the larger F will be. So, from your position on the wire you would see magnetic field lines flowing past you. You would find that the force F will depend on how fast these lines passed by. You could then write down an equation describing the electric field "induced" by this moving magnetic field, and if your name was Faraday you would have discovered "Faraday's law of induction:" an electric field is induced in any conductor moved through a magnetic field, or in any stationary conductor subjected to a moving magnetic field. The phenomenon of "induction" is basic to the operation of transformers and certain types of motors and generators.

Imagine pushing the wire in Fig. 6.10 back and forth through the magnet. If the ends of the wire are attached to a load, a closed electric circuit will be formed, and electrons will be pushed up and down the wire as the wire moves back and forth. This will produce an alternating current that flows first in one direction, then in the other, and so forth. The induced electric field in the wire will also oscillate between up and down, and so the voltage difference across the load will also oscillate between positive and negative values. The average values of voltage and current both will be zero (we'll examine the average power shortly). These oscillations in both current and voltage characterize an "alternating current" or "ac" circuit. Practically all the electrical power systems in operations today are of the ac type. The oscillation frequency in the United States is 60 hz (hz stands for "hertz," which is short for "cycles per second"), while in many other countries the 50-hz system is standard.

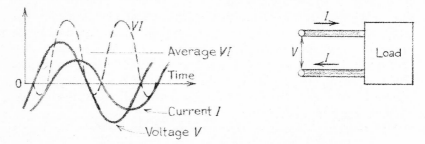

FIG. 6.12   VOLTAGE AND CURRENT IN AN AC SYSTEM.

Figure 6.12 shows the time histories of the voltage and current in an ac system. Note that the voltage and current do not necessarily change signs at the same time; i.e., they are "out of phase." So, the electric field might be tugging to the left while the current is still moving to the right. This means that one has to be careful in thinking about the work being done on the charges, because the motions and forces on the charges are not always in the same direction. Figure 6.12 also shows the time history of the product of voltage and current. This gives the instantaneous rate of energy flow past a point in the circuit, but because both V and I change sign, VI can be positive part of the time and negative the rest. In Fig. 6.12, positive VI corresponds to energy flow into the system, negative VI to energy flow out. If the current and voltage are 90° out of phase (meaning that one peaks while the other is crossing zero) as shown in Fig. 6.13a, the positive and negative contributions to the VI product will exactly cancel, and thus there will be no *average* energy flow in the circuit. In contrast, Fig. 6.13b shows a situation where V and I are "in

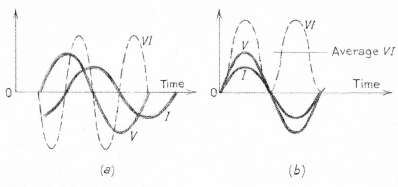

FIG. 6.13 POWER IN AC CIRCUITS.
(*a*) ZERO AVERAGE POWER,
(*b*) PURELY RESISTIVE LOAD.

phase" at all times, and hence the product *VI* is always positive. This means that the average value of *VI* is not zero, and energy is flowing into the system.

The phase relationship between *V* and *I* in an ac circuit is determined by the nature of the load. Resistive loads (Fig. 6.14*a*) draw currents that are in phase with the applied voltage, as in Fig. 6.13*b*. Common loads also contain "inductive" elements, such as coils in electromagnets. Do you remember the idea of "induced" electric fields? An oscillating current

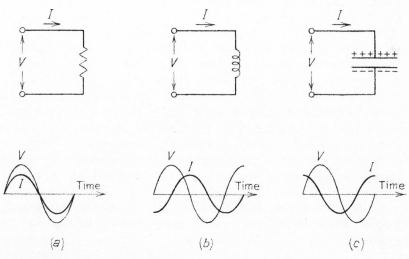

FIG. 6.14 TYPES OF AC LOADS.
(*a*) RESISTIVE, (*b*) INDUCTIVE,
(*c*) CAPACITIVE.

flow in a coil produces a changing magnetic field, which in turn induces an electric field in the coil. The induced voltage tends to fight off the applied voltage, which makes the current flow "lag" behind the voltage charge. In an ideal coil (Fig. 6.14b) the resistance is zero and the current flow is 90° behind the voltage, so that no power is expended; there are no I^2R losses ($R = 0$), and the process creates no entropy. So, ideally electromagnets do not require any power! However, real electromagnets do have some resistance, and so one must supply their I^2R losses.

Another type of load is a "capacitor," which is essentially a bucket for charge. Capacitors can be formed by separating two metallic plates by a nonconductor, which will not permit electrons to pass directly from one plate to another. As the voltage across the plates rises (Fig. 6.14c), electrons quickly flow to one plate, building up a negative charge. This pushes electrons off the opposite plate, which then acquires a positive charge. As the voltage difference reaches its peak, electrons start to rush away from the capacitor. So, in a capacitive load the current tends to "lead" the voltage (by 90° in a purely capacitive circuit).

We see that, although the average values of V and I are both zero in ac circuits, the average of the product VI need not be zero. In other words, although the charges are not really going anywhere, they have the capability of transmitting energy. This probably seems strange, and an analogy might be helpful. Think about waves on the ocean. A person in a raft bobs up and down, and doesn't really go anywhere as a result of the passage of waves. So, the waves don't really transport water; but they clearly transport energy, as any surfer knows. Ocean waves travel at speeds that depend in a complex way on the ocean depth and the wavelength of the wave. Electromagnetic waves travel at the speed of light, and hence travel from one end of a wire to another hundreds of miles away in a very small fraction of the period of one 60-hz oscillation. So, 60-hz electric waves are so long that there aren't any crests or troughs in the electromagnetic waves in a transmission wire. However, electromagentic waves at frequencies of 10^8 hz are much shorter; television transmissions use waves only a few feet long, the size of TV antennas.

How can we quantify the voltage and current in an ac system if the average voltage and current are both zero? We need measures of the amplitude of the voltage and current oscillations, and it is customary to use the "root-mean square" ("rms") values. The rms value of a periodic quantity like V is the square root of the average of the square of the quantity,

$$V_{rms} = \sqrt{(V^2)_{average}} \qquad (6.14)$$

For "sinusoidal" variations of the sort used in power circuits, the rms value is $1/\sqrt{2}$ of the peak value. So,

$$V_{rms} = \frac{V_{max}}{\sqrt{2}} \qquad\qquad (6.15a)$$

$$I_{rms} = \frac{I_{max}}{\sqrt{2}} \qquad\qquad (6.15b)$$

where V_{max} and I_{max} are the maximum voltage and current that occur during the cyclic alternation. Ac voltmeters generally read V_{rms} and not V_{max}, and the 110-volt rating on household systems refers to V_{rms}.

The average energy flow in an alternating system can be expressed in terms of the rms voltage and current. For a purely resistive load, such as light bulbs and toasters,

$$P = V_{rms}I_{rms}$$

For other loads the power is expressed as

$$P = \alpha V_{rms}I_{rms} \qquad\qquad (6.16)$$

where α is the "power factor" determined by the phase relationship between voltage and current discussed above. The product $V_{rms}I_{rms}$ is called the "apparent power," and is usually expressed in kilovolt-amps (kva) rather than in kilowatts to avoid confusion with the true power given by Eq. (6.16). Power lines and large transformers are normally rated in kva. The power factor is equal to 1 for a purely resistive load, and otherwise is less than 1. Recall that, for inductive loads the current tends to peak somewhat after the voltage, and one says that the "current lags the voltage" by some angle ϕ. In contrast, in capacitive loads the voltage peaks somewhat after the current, and one says that the "current leads the voltage." The power factor α turns out to be just the cosine of the angle ϕ. If the power factor is low, either V_{rms} or I_{rms} will have to be large for a given power, and thus transmission-line engineers like to have power factors very close to unity to help keep transmission losses down. The power factor of a load is sometimes purposely changed by adding either induction coils or capacitors to the circuit at appropriate points in the distribution system; since most loads are inductive, capacitors are often used in transmission-line substations for power-factor correction. Power factors range from 0.7 to 1 in typical operations.

The resistance of a transmission line to electric-current flow produces I^2R line loss in both ac and dc circuits. As with dc circuits, ac transmission lines are designed for high voltages to minimized line losses. The common scheme is to generate energy at relatively low voltages (perhaps 1,000 volts), transform this to high voltage for transmission, and then transform down to lower voltages for distribution at "substations" in urban areas. The final step down to 110 volts generally occurs very close to the load, and you probably have noticed the transformers that do this job at the top of power poles near your home.

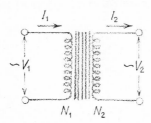

FIG. 6.15 TRANSFORMING AC POWER.

Transformers are basically coils of wire that are not connected electrically but are coupled magnetically. The changing current flow in the primary coil induces a changing magnetic field in the secondary coil, and this changing magnetic field in turn produces a changing electric field in the secondary coil by the induction process. The electric and magnetic fields in the secondary coil act on electrons in the wires, and if the circuit is closed, an alternating-current flow is set up in the secondary circuit. The equations of electromagnetics reveal that, if the coils are closely coupled by the alternating magnetic field that they share, the voltage across each coil is proportional to the number of turns in the coil. So, if N_1 and N_2 are the number of turns in the two coils, and V_1 and V_2 are the (rms) voltages across the coils (Fig. 6.15), then,

$$\frac{V_1}{V_2} = \frac{N_1}{N_2} \tag{6.17}$$

For an ideal transformer, the power input to the primary coil all appears as the output of the secondary coil. Hence, if the primary voltage is *greater* than the secondary voltage, the primary current must be *less* than the secondary current. Hence, in an ideal transformer the currents in the two coils must be inversely proportional to the number of turns,

$$\frac{I_1}{I_2} = \frac{N_2}{N_1} \tag{6.18}$$

Transformers designed to operate at high voltages have trouble with arcing between windings of the coils, and special designs are required to eliminate this problem. Real transformers have some I^2R losses, of the order of a few percent of the transmitted power. The electrical energy dissipated in I^2R heating must be removed to permit steady-state operation. Small transformers, such as those in radios and TV sets, are cooled just by the natural circulation of air. Larger transformers used in power-transmission networks are usually oil-filled and sealed; the oil assists the natural convection process, and also helps prevent arcing. Very large transformers require forced cooling with air or water.

The oscillations in power in ac circuits mean that energy does not flow steadily from the source. This represents an inefficiency in the use of the generating and transmitting system, and so electrical engineers devised the "three-phase" system to eliminate this problem. The idea is to use three companion ac circuits as shown in Fig. 6.16. Voltages (relative to ground) in the three wires are "120° out of phase," meaning that they peak at different times in the cycle of the basic oscillation. So, when the instantaneous energy flow to load 1 is dropping, the energy flow to the other two loads can be rising, and if the average power used by each load is the same ("balanced loads"), the instantaneous *total* power outflow from the source and down the transmission system will be constant.

There are two basic schemes for connecting three-phase power. In the Δ-connected system of Fig. 6.16a each load sees the full rms line-to-line voltage, and the current coming down each line splits to pass through two loads. In the Y-connected system of Fig. 6.16b the full line current is fed through each load, and the rms voltage across each load is less than the rms voltage between two lines. In the United States practically all electrical energy is generated and transmitted as three-phase power. You will notice the three lines high atop towers in the high-voltage transmission lines.

Almost all industrial users require three-phase power for motors, but

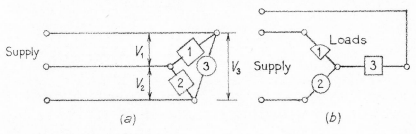

(a) (b)

FIG. 6.16 THREE-PHASE POWER.
(a) DELTA CONNECTION, (b) Y
CONNECTION.

most homeowners require only single-phase power. So, at substations the high-voltage power is transformed to some intermediate voltage for transmission within the city, still as three-phase power. At thousands of locations within the city transformers are hooked up to *two* of the three wires on the power pole, and low-voltage *single*-phase power is delivered from the transformer to houses nearby. The utility company tries to balance the load seen by the three-phase supply line by keeping about the same number of houses supplied by each of the three pairs of wires in the three-wire three-phase line. Figure 6.17 shows a schematic of this sort of system. Note that two houses have only the normal 110-volt single-phase supply but that *three* wires lead from the single-phase transformer to one of the houses. The central wire is a "ground" wire, and the other two wires are at an rms voltage of 110 volts above the ground. However, the rms voltage difference between these two "hot" lines is 220 volts. The 110-volt single-phase system is used for lighting and small appliances, and the 220-volt single-phase supply is used for stoves, clothes dryers, and shop motors. The more ambitious homeowner-craftsman might want to have three-phase 220-volt power in his garage, which the utility company will gladly supply for a price.

Actual electrical energy distribution systems are much more complicated than Fig. 6.17 suggests. There may be additional transformation

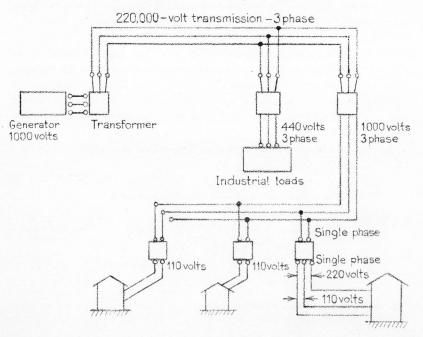

FIG. 6.17 SCHEMATIC OF PUBLIC ELECTRICAL POWER DISTRIBUTION SYSTEMS.

stages, switches, and circuit breakers, and the outputs of several power stations will usually be connected so that they jointly supply all the loads. Whenever the power required by the various loads changes, as, for example, when you turn off a light, somewhere a generator will have to cut back slightly on its power output. If a major industrial user cuts back suddenly, and if the generator can't respond quickly enough, the voltages in the system can build up quickly, triggering circuit breakers, shutting off other loads, and thereby causing all kinds of havoc. Or, if one power plant shuts down suddenly from a malfunction, others have to take up the load, and if they can't respond quickly enough, the system voltages may drop, lights dim, and motors stall and then stop from excess current when the system voltage comes up suddenly. Peculiar chains of events such as these have produced sudden "blackouts" over wide areas. There are many advantages to having such a complex "electrical grid" system, but it does couple many users to one another and many suppliers to many others. But this is true of many things in any complex society.

GENERATORS AND MOTORS

Let's now look at how generators make electrical energy and how motors convert this to useful work. Electrical generators consist of a magnetic field and a complex of moving wires called the "armature." Figure 6.18 shows a schematic that is useful for understanding how they

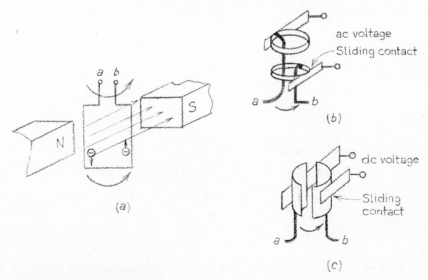

FIG. 6.18 SCHEMATIC OF GENERA-TION PRINCIPLES. (a) BASIC ARMATURE, (b) SLIP RING CONNECTOR, (c) COMMUTATOR.

work. As the armature is turned through the magnetic field, a potential difference develops at the ends of the armature wires by the induction processes described previously.* For half the rotation cycle the voltage difference between points *a* and *b* will be positive, and for half it will be negative. So, if terminals *a* and *b* are connected to an external circuit through "slip rings" as shown in Fig. 6.18*b*, an alternating current will flow in the external circuit. However, if "commutators" are used (Fig. 6.18*c*), the external circuit will always see a voltage of the same sign, and hence a direct current will flow in the external circuit. So, in Fig. 6.18 we see the basic ideas behind both ac and dc generators.

Of course, in practice things are more complicated. Figure 6.19 shows more detail on the dc "shunt-wound" generator, which uses energy drawn from the generator output to provide current to an electromagnet that provides the magnetic field. The copper field windings have such a low resistance that a large current will flow in the field windings unless an additional resistance is placed in the field circuit, and the shunt resistor is therefore added to control the field current. The generator is normally started without any external load, and the armature current is used solely to power the field magnet. As the field builds up in strength, the output voltage increases, and this further increases the field current and magnetic field strength. The armature voltage therefore continues to build up, but not indefinitely. The coils of the field magnet are wrapped around an iron core, and when this core becomes fully magnetized, the field strength stops increasing. The output voltage therefore builds up until a steady-

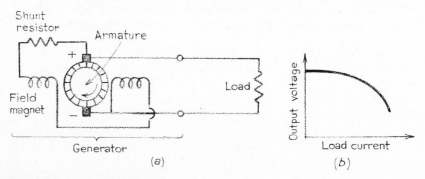

FIG. 6.19 SHUNT-WOUND DC GENERATOR. (*a*) SCHEMATIC, (*b*) CHARACTERISTICS.

* Remember, if you take a reference frame on the stationary magnet, you explain the voltage differences as resulting from the forces that the magnetic field exerts on the electrons in the moving wire. If instead you ride on the armature, you explain the voltage difference as a result of the moving magnetic field that you see from this reference.

state level is reached. The no-load voltage can be controlled by adjusting the shunt resistance, which in turn alters the field current and hence the strength of the field magnet. So, it is easy to control the output voltage from a dc generator.

Suppose now we connect a load to the generator; you can think of the load as a bunch of light bulbs. This will permit electrons to move freely out of the generator and through the load circuit back around to the generator. The armature voltage will drop slightly, and so the strength of the field magnet will be reduced somewhat. But the shunt resistance can be reduced to increase the field current so as to keep the output voltage constant. The generator is now putting out electrical energy that it must be getting from somewhere. The motor, turbine, or horse driving the generator must be working hard to drive the generator in order to force electrons to move through the armature and load.

If the shunt resistance is not changed by hand or automatically as the load is increased, the output voltage will fall off for the reasons explained above, and so a dc shunt-wound generator will have the output characteristics shown in Fig. 6.19b. A clever bit of engineering gets around the drooping-voltage problem. By placing another winding in the field, and passing the full armature current through this field, the strength of this "series" field winding will *increase* as the armature current goes up, and this provides a tendency for the output voltage to *increase* with increasing load. By proper balancing of the series and shunt windings, a machine can be made in which the output voltage is nearly *independent* of load, as shown in Fig. 6.20. These generators are called "compound" dc generators; they are chiefly used in situations where the load is subject to rapid fluctuations, as, for example, in an urban rapid-transit system.

Generators for high current and voltage output have difficulty with

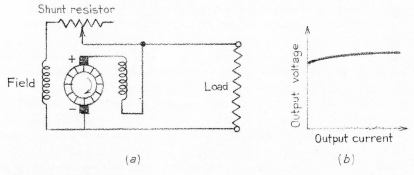

FIG. 6.20 COMPOUND DC GENERATOR. (*a*) SCHEMATIC, (*b*) CHARACTERISTICS.

arcing in slip rings or commutators. This can be eliminated by turning the generator "inside out," i.e., by rotating the magnetic field while holding the armature windings stationary. The magnetic field is produced by electromagnets mounted on the internal rotor, but these require relatively small amounts of current that can easily be handled with slip rings and carbon or copper mesh brushes. The armature (or "stator") consists of an iron ring with induction coils wound on the inside. The magnetic poles on the rotor move past the ends of these coils at a very short distance from them. Since the current in the armature loop will alternate in direction, as described previously for the "inside-in" generator, this scheme is suitable only for generation of alternating current. The frequency of alternation is determined by the rotor speed and the number of poles. Practically all large generators deliver three-phase power, which can be obtained by proper angular spacing of the windings on the armature and stator. In power station applications the speed of the generator is accurately controlled to give precisely 60-hz oscillation. The output voltage is then fixed by the generator design but may be trimmed by altering the current supplied to the rotor. Typical machines operate at 550, 1,100, 2,200, 6,600, 13,200, and 20,000 volts.

Electric motors are essentially the opposite of generators. Since magnetic fields produce forces on moving electrons, if one forces a current to flow in the armature windings on a rotor the magnetic forces on the moving electrons will be transmitted to the wires, and with proper configuration the rotor can be made to turn. In a dc motor the current is supplied through commutators to the armature. The motor speed for a given load will depend upon the magnetic field strength and the armature current, either of which can be adjusted to control the speed. The most common approach is to vary the dc supply voltage, which allows one to adjust the motor speed over a wide range. This feature of dc motors makes them particularly useful in applications where speed control is important.

The resistance of the armature windings is generally very low, and so they tend to draw a lot of current when not moving. Once the rotation begins, the moving armature windings act like a generator and develop a voltage that tends to oppose the current driving the motor, holding the armature current at a tolerable value. So, start-up of dc motors presents a problem; in order to limit the armature current on a start-up, a resistance must be placed in series with the armature. This must be removed as the machine comes up to speed. Automatic motor-starter boxes are usually used for this purpose. If you have ever had the thrill of a ride on an electric trolley, you may remember the operator using a crank handle to get things going. This is how he alters the armature series resistance to get the trolley moving. There are many other subtleties of winding that are used in dc motors, each being engineered into the motor to help in starting, speed control, smoothness of load change, efficiency, etc. But the basic ideas are the simple ones outlined above.

The most common type of ac motor is the induction motor. It consists ordinarily of a stator to which power is supplied. The alternating-current flow in the stator produces a rotating magnetic field that is used to pull around the iron core of the rotor. In "squirrel-cage" induction motors the iron core of the rotor has slots in which run heavy copper bars parallel to the rotation axis. These conducting bars are connected at the ends by copper rings, and it is this cagelike structure of the copper conducting circuit that gives the motor its name. An alternating current flows in the rotor, which is not connected electrically to external circuits. The rotor current acts to magnetize the rotor, and thus provides a push against the rotating magnetic field in the stator. The rotor conductors must cut the rotating magnetic field lines in order to develop current, and hence the rotor must turn at a slightly lower speed than the rotating magnetic field. The amount of "slip" depends upon the winding scheme and the spacing of rotor and stator; many motors operate at 1,750 rpm with a rotating magnetic field that spins at 1,800 rpm when connected to a 60-hz ac supply. Three-phase induction motors spin in the direction of the rotating magnetic field, and this direction can be reversed by interchanging two of the connecting wires. Single-phase induction motors run in the direction that they are started, and this can be controlled by special "shading coils" that are wound into the machine to produce a weak preferential rotation for starting. These may require current through brushes and slip rings, but only small amounts of current are required, and thus the sliding contacts do not present any serious engineering problems.

Squirrel-cage induction motors tend to produce high torque at near-maximum speed, and have low torque at low speed. This is not desirable for applications requiring high torque at start-up; for such situations "wound-rotor" induction motors are preferable. The squirrel cage is replaced by windings that are connected to an adjustable resistance outside the motor through slip rings. By adjusting the resistance, the rotor current can be controlled, and this permits control of the motor torque-speed characteristics. Figure 6.21 shows the torque-speed characteristics of typical squirrel-cage and wound-rotor induction motors.

Three-phase induction motors will run if only two lead wires are connected to a single-phase power supply. However, three-phase voltages are generated by induction in the stator windings, and hence the machine can be used to convert single-phase power to three-phase. The motor

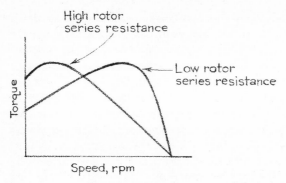

FIG. 6.21 CHARACTERISTICS OF AC INDUCTION MOTORS.

runs without shaft power output, and is then termed a "phase converter." Railway locomotives often use phase converters, which permit the track to supply only two-phase power and the drive motor to operate on three-phase power.

The speed of an ac motor is virtually independent of the supply voltage, and can be varied only by changing the frequency of the ac supply, by altering the winding, or by changing the slip. Some motors employ two sets of windings and hence can run at two speeds. The speed will of course drop off slightly as the motor is loaded up; i.e., the slip increases slightly with load. For many applications it is desirable to have a motor that runs at precisely a constant speed, independent of load. The "synchronous motor" does this by using a dc power supply to provide the rotor current, in which case the rotor turns at *exactly* the rotation rate of the stator magnetic field. The dc supply is often made an integral part of the motor. Clock motors are of this type.

The efficiency of electric generators and motors depends upon the design, load, speed, power factor, and other operating factors. Large three-phase generators have efficiencies (conversion of mechanical to electrical), as high as 98%. Small (1-hp) motors may have part-load efficiencies as low as 75%, while large three-phase motors have efficiencies as high as 95%. Generator efficiencies tend to be lower at low power factor, which is why so much importance is placed on keeping the power factor of the total load on an electrical power network as close to unity as possible. The mechanical energy that is not converted to electrical energy in a generator, and the electrical energy that is not converted to mechanical energy in a motor, of course becomes thermal energy that must be removed from the machine. In large machines this constitutes considerable power, and so elaborate cooling provisions are often employed.

DIRECT-ENERGY-CONVERSION SYSTEMS

In previous chapters we described the technology of conventional power-generating systems. We saw the long chain of energy conversion from chemical to thermal in a combustion chamber, thermal to mechanical in a working fluid cycle, and then finally mechanical to electrical in generators as just discussed. There is considerable interest in schemes which avoid some or all of the intermediate stages in this process, and since the end of World War II there has been considerable research and development on various schemes for "direct energy conversion." The fuel cell discussed in the last chapter is basically a device for going directly from chemical energy to electrical energy. In this section we will look at some important schemes for going more directly from thermal energy to electrical energy, schemes that do not employ a conventional electrical generator.

The first "DEC" device to be used to any extent was the "solar cell," a device for conversion of incident solar energy directly into electricity. Solar cells today are used on practically all space vehicles, on transistor radios for the beach, and in exposure meters on cameras. Light particles ("photons") striking the molecules in a semiconductor (such as cuprous oxide) dislodge electrons, which move easily into an adjacent metallic conductor (such as copper). This gives the copper a negative potential with respect to the semiconductor, and hence electrons will flow through an external circuit to return to the semiconductor (Fig. 6.22). The voltage is determined by the materials used; practical cells have voltages of the order of 1 volt. The power output is limited by the incident solar energy and by the efficiency of the cell; modern cells are able to convert only about 15% of the incident solar energy to electrical energy, which means that vast areas are required for generation of large amounts of

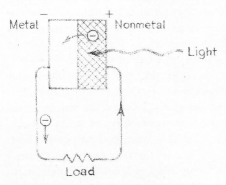

FIG. 6.22 THE PHOTOVOLTAIC CELL.

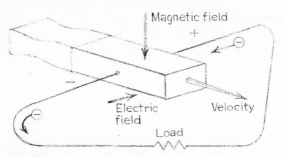

FIG. 6.23 THE MHD SYSTEM.

power. Schemes for doing this in orbit and sending the electrical energy down to earth in a microwave beam have been analyzed in some detail, but this is a bit far-fetched for implementation in this century.

A DEC scheme that will be implemented in the very near future goes by the mindboggler "magnetohydrodynamics," or "MHD" for short. The MHD system relies on the basic phenomena of electromagnetics, namely, that a voltage will develop along an electrical conductor moved properly through a magnetic field (hence "magneto"). In a conventional generator the conductor is a copper wire, but in an MHD generator the conductor is a fluid (hence "hydro") moving at high speed (hence "dynamic"). The MHD system is shown schematically in Fig. 6.23. The fluid is a gas at a very high temperature (2000 to 5000°K), "seeded" with elements such as cesium that ionize (lose electrons) at these elevated temperatures. The presence of free electrons in the gas makes it capable of conducting electric current. The gas is accelerated to a high velocity in a nozzle, usually a supersonic nozzle of the rocket type, and then sent through a strong magnetic field. The magnetic field pushes the moving electrons sideways in the flow channel, causing a voltage difference to develop between the two sides of the channel. Electrons will flow around through the external load when these sides are connected to an external circuit. Thus, the MHD generator can be used to produce dc power, which of course can be converted to ac outside the MHD system in a variety of ways. The high-temperature gas can be produced by combustion of fossil fuels, and the MHD system thereby eliminates the turbine, heat exchangers, etc. The real advantage is the possibility of using temperatures well above those that can be used in a conventional power plant, and this means that the potential efficiency of conversion from thermal energy to useful electrical energy is higher than in a conventional system, at least so says the second law of thermodynamics.

There are many practical problems associated with MHD generators.

The surfaces used to draw off the electric current ("electrodes") must not get too hot or they will melt; so, one thinks of cooling them with water. But if the gas close to the electrodes gets too cold, it will no longer be ionized, and hence it will not conduct the desired current. Potential differences also tend to develop along the channel in the flow direction, and this gives rise to currents flowing parallel with the flow. So, one thinks of segmenting the electrodes. But this gives rise to other problems, and on and on it goes. Nevertheless, the promise is so great that considerable research and development effort has been devoted to MHD, particularly in the Soviet Union. One of the major problems is that the magnets used in experimental MHD channels require large amounts of power, usually more than the channel itself puts out! One must go to very large sizes to get a system in which the MHD channel output power is enough to drive the magnets and still have a net power output.

The problem of high magnet power will very likely be solved through use of "superconducting magnets." At very low temperatures (e.g., 4°K) certain materials become "superconductors," in that they offer essentially *zero* resistance to the flow of electric current. So, by winding a magnet out of such a material and cooling it to superconducting temperatures, the magnet-power requirements can be made virtually negligible. Of course, one must consider the power required to maintain the magnet at the low temperature; with perfect insulation against heat loss one could simply immerse the magnet in liquid helium, and space-age technology has given us insulations that are remarkably close to perfect. So, the technology for using superconducting magnets is available, and at this writing the first small experimental MHD channel to operate with a superconducting magnet was being fired up in the High Temperature Gasdynamics Laboratory at Stanford University.

Many practical problems with MHD power systems remain to be solved, others to be discovered. The best way to couple MHD in with conventional power systems is still a subject of study. It is not possible to extract all the energy from the hot gases using MHD. As energy is extracted, the gases cool, and this reduces their ability to conduct electricity. So rather than throw away the energy in the moderately hot exhaust from an MHD generator, one can use this energy to boil steam for a conventional steam power system. The MHD unit then becomes a "topping cycle," as indicated in Fig. 6.24. Figure 6.25 shows the world's first MHD-steam power station designed along these lines during its construction in the Soviet Union, where a considerable research and development program in MHD has been going on for over 20 years. As experience is obtained with this pilot system, the problems of MHD will gradually be solved; it is quite likely that there will be several

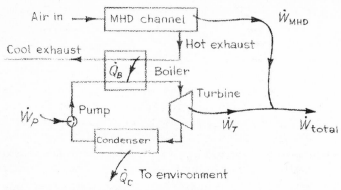

FIG. 6.24 MHD-STEAM BINARY SYSTEM.

MHD-conventional power systems in operation in both Russia and the United States by the end of this century.

Another DEC system that shows promise for applications at lower power levels is the "thermoelectric" converter. This device takes advantage of the fact that a voltage difference develops in a circuit composed of two dissimilar conductors or semiconductors subjected to temperature differences as shown in Fig. 6.26. Thermoelectric phenomena provide a means for direct conversion of heat into electrical energy, and a number of small devices using this effect have been built. In the Soviet Union kerosene lamps are used to provide the heat for thermoelectric generation

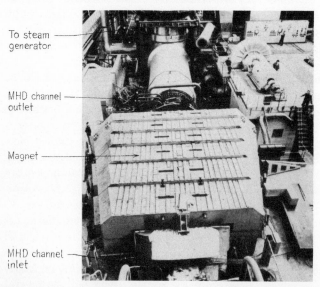

FIG. 6.25 MHD PLANT UNDER CONSTRUCTION IN THE SOVIET UNION.

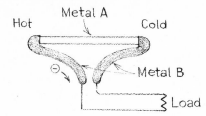

FIG. 6.26 THERMOELECTRIC CIRCUIT.

of electricity to operate small transistor radios in rural areas. The efficiency of thermoelectric generators is very low (10%), and because they are inherently irreversible devices, there is no hope of approaching the Carnot efficiency. Thus, the facts of life and science have limited the development of thermoelectric technology. Although thermoelectric systems can never be fully "reversible," they can be operated in reverse; i.e., a current can be supplied with the result that heat is pumped from a low-temperature region to a region at higher temperature. If you enjoy reading patents, read number 3080723, which describes a thermoelectric blanket that (in principle) can be used for either heating or cooling; this device has never been marketed, probably because of the immense practical problems involved.

In spite of their low efficiency, the compactness and mechanical simplicity of thermoelectric converters make them especially well suited to applications where long operation without servicing is required. The first nuclear power system to operate in space involved a "radioisotope" source of thermal energy and a thermoelectric converter. In these systems radiation from the decay of radioactive materials collides with atoms, exciting them to vibrate in a random manner, and hence increasing the internal energy of the system. This makes it hot, and energy can be taken out of the system as heat to supply a thermoelectric generator. The generator must reject some "waste heat" at a lower temperature, and so must be provided with cooling. Systems of this sort were used on Apollo missions to the moon and other space missions. The French have developed a small radioisotope-powered thermoelectric converter for implantation in the body. This device generates electrical pulses for stimulation of the heart, and is in effect a "nuclear thermoelectric pacemaker." The amount of power delivered is very small, about 40 microjoules per pulse at 72 pulses per minute. Thermoelectric generators will probably be used widely some day in very low power applications of this sort.

The "thermionic converter" works on the phenomenon of "thermionic emission," the streaming of electrons from a very hot metallic surface. In vacuum tubes used in radio and television sets this principle is used to generate a stream of electrons that are then controlled to perform the desired function of the tube. In the thermionic converter "diode" the electrons are collected on a cold surface in the "tube" and then allowed to flow around through an external load circuit back to the emitting surface. The emitter, or "cathode," assumes a positive potential, and the collector, or "anode," is the negative terminal of the converter. Energy must be transferred from the anode as heat in order to maintain it at a suitably lower temperature. The efficiency of the generator is limited by the Carnot efficiency, and the generator can be viewed as a heat engine using electrons as the working fluid. Figure 6.27 shows the basic idea of a thermionic converter. Thermionic-converter diodes are sometimes vacuum tubes, but better performance has been obtained with gas-filled tubes.

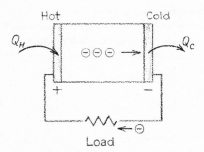

FIG. 6.27 THERMIONIC CONVERTER.

Vacuum diodes must have very close spacings (of the order of 0.02 mm), for otherwise the retarding potential arising from the distribution of electrons in the gap becomes too great for satisfactory operation. Diodes of this type have been operated with cathode temperatures of the order of 1500°K and with the anode at approximately 900°K, yielding power densities of the order of 2 to 10 watts/cm^2. The energy-conversion efficiency of these systems is of the order of 3 to 12%. The small spacings are difficult to maintain, especially with large cathode-anode temperature differences, and short-outs are frequent. The gas (plasma) diodes employ easily ionizable elements, such as cesium, whose positively charged ions tend to neutralize the retarding effects of the space charge, allowing larger spacings to be employed (1 mm). Efficiencies of the order of 15% have been obtained with this type of plasma diode operated at cathode temperatures of about 2500°K.

The thermionic converter is well suited for the production of electrical power in space from a nuclear-heat source. Since the energy which must be rejected as heat must be radiated to space, and high temperatures are required for low-weight radiators, space-systems designers can accept low efficiency in the interests of lightweight nonmechanical power-conversion equipment. The thermionic converter will probably also find application as a topping device for Rankine power systems.

PROBLEMS

For getting a feel for numbers

6.1 A 110-volt household circuit is fused at 30 amps. What is the maximum power that can be drawn from this circuit?

6.2 A house has 10 of the circuits described in Prob. 6.1, and in addition has two 220-volt circuits fused at 40 amps for the stove and electric dryer. What is the maximum power that can be consumed in this house?

6.3 A small factory has 440-volt wiring capable of handling 50 amps. The power factor for the plant averages around 0.85. What is the maximum electrical power that can be used in the plant?

6.4 A farmer wants to run dc power from his barn to a pump 2 miles away. He has plenty of 14-gauge copper wire (0.064 in diameter) but is worried about the voltage drop in lines 2 miles long. The pump will draw a current of 20 amps when supplied with 80 volts, and will operate properly when so supplied. In order to make up for voltage drop in the current-carrying lines, the voltage at the barn must be higher than 80 volts. Compute the required voltage difference between the lines at the barn, the pump power, and the I^2R power loss in the two lines.

6.5 A three-phase high-voltage line carries power 200 miles from the generating station to the city. The line current is 1 amp and the "apparent power" is 220 kva. Determine the wire diameter that must be used if the I^2R line losses are to be kept below 10% of the apparent power.

For getting a feel for hardware

6.6 Take a trip around your local urban area. Locate the power-distribution stations, and find out what voltages are input and output from this station (usually posted on the poles). Look at the pole transformers in your neighborhood; are they two-phase or three-phase units (how many input and output wires?) Does your local pole carry more than the common 110-volt household supply? What other services are carried on your local poles? Who owns the poles? Write a brief account of what you have learned about the distribution system that serves you electricity.

6.7 Visit your local college library, and scan through some recent copies of *Electrical Review*, *Electrical World*, and *Power Engineering*. Look at the hardware shown in the advertisements, and try to figure out what it is for. Try reading some of the descriptive articles. Write a brief account of your impression of the state-of-the-art in the field of electrical power and the sort of things that are currently of interest in the field.

6.8 Unplug your toaster, and remove the cleanout plate so that you can inspect the heating element. Why does it get hot? Is it made of different material than its supports and the wires that supply it? Why? Carefully break open a light bulb and inspect its internal design. Why is it built the way it is? Unplug your electric clock and carefully remove the case. What does it say on the motor? Does the motor run slower when the lights dim? Write a brief account of what you have learned about the electrical devices that you use at home.

6.9 Climb up into your attic and explore the myriad of wires that supply electrical energy to various plugs and lights. Is there anything special about the doorbell system? Switch off the circuit breakers or remove the fuses in your house or apartment one at a time. Determine how your place is wired. Are the lights and plugs on the same circuits? Do you have more plug circuits or light circuits? How much power can you draw on each circuit? Write a brief account describing what you have learned about the distribution of electricity in your home.

6.10 Look over the electrical wiring in your automobile. What is the primary supply voltage? Is it ac or dc? Check the bulbs; does the same current flow through any two bulbs (does one go out when another is removed)? Where are the switches that control the various bulbs? What sort of repair conveniences are employed in the wiring? What sort of electric motors and generators does your car employ? Your battery is marked with its capacity in "ampere hours." Knowing the voltage, you can compute the total energy available from your battery. Compute the number of equivalent "engine seconds" at the advertised horsepower of your engine. Check the fuses in the electric circuit, and use this to estimate the current to your headlamps. Compute their power consumption in watts, and check this against any printed ratings. How long could your battery supply your headlights without recharging? Write a short description of the electrical energy system in your automobile and the things in it that you now feel competent to fix yourself.

To increase your independence

6.11 Borrow a copy of the "Mechanical Engineer's Handbook," and read the sections on power transmissions and distribution in the chapter on Electrical Engineering. Note in particular the material on the economics of different sorts of transmission systems. Then, visit your local utility company and find out the details of their distribution system. Try to rationalize differences between the handbook recommendations and the realities of the local system. Write a short description and analysis of your local distribution system.

6.12 Read New Superconductors in *Scientific American*, November 1971, and Superconductors for Power Transmission in *Scientific American*, April 1972. Contact one of the organizations cited as being active in the development of this technology, and find out where they stand. Then, write a short description of the idea and the outlook for the development of superconducting transmission systems.

6.13 Visit a factory that uses heavy machinery. Inspect the nameplates on the motors and generators that they use. Where do they use dc? How is the speed of the machines controlled? What voltages are available within the plant? What is the daily consumption of electrical energy? Estimate the energy used for lighting each day.

Contact the utility company that supplies this plant, and find out the rate at which they are charged for electrical energy. Compare this with the rate for homeowners, and rationalize any difference. Write a short paper on the utilization of electrical energy in this plant. What would they do if a federal order required them to reduce their consumption of electrical energy by 10%? What does this experience suggest to you about the feasibility of a national policy to effect electrical energy conservation?

6.14 Go to your local college library and scan the stack material on electrical energy. You will very likely find some interesting information on the electrical energy distribution systems in various countries. Write a paper summarizing the history and current status of the electrical energy "grid" in a country of your choice. How do the cultural and political values of the country impact upon the distribution of electrical energy in that country?

6.15 Look up some recent articles on direct-energy-conversion devices in technical magazines and journals in your local library. Choose one DEC device, and write a short paper tracing its development, current state-of-the-art, current activity, and the outlook for its utilization in the next 20 years. What sorts of breakthroughs in the technology of the device might change the outlook dramatically? What "conventional" technology would be likely to be most seriously upset by such a breakthrough? Are there potential uses of the device that might be regarded as "weaponry" that have either accelerated or impeded the development of the device?

7

ATOMS WORK HARDER

In which we see
the power of
twentieth century alchemy

NUCLEAR ENERGETICS

In this chapter we will examine the science and technology of nuclear energy. Let's begin by reviewing the basic picture of the atom and some of the things that happen in the atomic world. Every atom consists of a "nucleus," the very dense central core, and "electrons" that swirl something like planets around the nucleus. Electrons are very light particles with a negative charge; the number of electrons varies from element to element, and this is what makes the elements different. The nucleus contains positively charged particles called "protons"; although a proton has the same magnitude of charge (but opposite sign) as an electron, the proton is much heavier. The numbers of protons and electrons are identical, and thus atoms are electrically neutral. The nucleus also contains some electrically neutral particles called "neutrons" that have about the same mass as protons. There are about the same number of neutrons in the nucleus as protons. Two atoms with the same number of electrons (and hence protons) but different numbers of neutrons are called "isotopes" of the same element. Since chemical reactions involve only electrons, all the isotopes of a given element behave the same way chemically. But they behave quite differently when nuclear reactions are involved, and that is what this chapter is all about.

The masses of the fundamental particles mentioned above have been accurately measured; in terms of the atomic mass scale, which defines the mass of the carbon atom $_6C^{12}$ to be 12.0000 amu ("atomic mass units"), these masses are

Neutron	1.008665 amu
Proton	1.007825 amu
Electron	0.00055 amu

One amu is equivalent to 1.66×10^{-24} grams, and so you can see that these particles are all very, very tiny. Let's check out the $_6C^{12}$ atom, which contain 6 protons, 6 neutrons (hence the mass number is 12), and of course 6 electrons:

Mass of 6 protons	=	6.04695
Mass of 6 neutrons	=	6.05199
Mass of 6 electrons	=	0.00330
Total		12.10224

Why does this not add up to 12.00000, as advertised? The reason is that the mass of an object, which you remember measures its resistance to acceleration, is not quite conserved. In other words, when you bring the particles together to form the atom, the mass of the whole is not quite the same as the mass of the parts. However, a certain amount of energy is released when the protons, neutrons, and electrons come together to form the atom, and thanks to Dr. Einstein we can think of this energy as taking away the mass; some of the mass was converted to energy. The theory of relativity proposed by Einstein and subsequently tested experimentally states that the energy E made available by the disappearance of mass m is related to the speed of light by the famous equation

$$E = mc^2 \tag{7.1}$$

Let's use this to compute the energy release upon formation of the $_6C^{12}$ atom, which is called the "binding energy" of the "nuclide" (another term for isotope). The mass defect is

$$12.10224 - 12.00000 = 0.10224 \text{ amu}$$

It will be convenient to find the energy equivalent of 1 amu. Since the speed of light c is, in the cgs unit system,

$$c = 3 \times 10^{10} \text{ cm/sec}$$

the energy equivalent of 1 amu = 1.66×10^{-24} grams is

$$1.66 \times 10^{-24} \text{ grams} \times (3 \times 10^{10} \text{ cm/sec})^2$$
$$= 1.49 \times 10^{-3} \text{ gram-cm}^2/\text{sec}^2$$

This will not mean much if you are not calibrated to energy in gram-cm^2/sec^2. It might help to know that "erg" is short for "gram-cm^2/sec^2," and hence the energy equivalent of 1 amu is 1.49×10^{-3} erg. About 10^{10} ergs make a Btu, and so the binding energy of a carbon nucleus is of the order of 0.15×10^{-13} Btu ... not very much by comparison with your daily needs. However, a thimbleful of carbon contains of the order of 10^{22} atoms, having a total binding energy of the order of 10^7 Btu! So, the energy available from nuclear reactions is potentially enormous.

In high-energy physics another energy unit is in very common use. This is the "electron volt" (ev), which is equivalent to the energy acquired by any particle carrying one electron's worth of charge when it falls through a potential of one volt. One electron volt is equivalent to

$$1 \text{ ev} = 1.6 \times 10^{-12} \text{ erg}$$

Since an erg is about 10^{-10} Btu, an electron volt is about 1.6×10^{-22} Btu ... not much at all, but then again there are an enormous number of atoms here and there. In terms of electron volts, the energy equivalent of one atomic mass unit is

$$1 \text{ amu} = \frac{1.49 \times 10^{-3} \text{ erg}}{1.6 \times 10^{-12} \text{ erg/ev}} = 0.931 \times 10^9 \text{ ev}$$

The notation "Mev" (million electron volts) is used, and so

$$1 \text{ amu} = 931 \text{ Mev}$$

The term Bev (billion electron volts) is used too, and you might easily remember that 1 amu of mass is roughly equivalent to 1 Bev of energy. So, when you read about a particle accelerator that produces 1-Bev particles, you should realize that the energy equivalent of the mass of an entire neutron is pumped into each accelerated particle.

In the "normal" atoms the electrons are well behaved, each whirling about in its proper orbit. However, collisions between atoms, even the sort that arise when an element is heated in a flame, can "excite" one or more electrons to higher than normal energy levels. One can think of the electron "shifting to a higher orbit" as a result of taking on the extra energy. Atoms in these excited states tend to be unstable, and the electrons eventually fall back to their normal orbit. When they do, the energy difference between the excited and normal level is emitted as electromagnetic radiation, which can be thought of either as "waves" or as a "photon particle." The frequency of the radiation is related to the energy through "Planck's equation,"

$$E = h\nu \tag{7.2}$$

where h is Planck's constant, 6.62×10^{-27} erg-sec, and ν (the Greek letter "nu") is the frequency of the radiation (cycles/sec). This has all been deduced by physicists as a means for explaining the peculiar light-emission characteristics of heated globs of different elements. Physicists use measurements of the frequency to calculate the energy jump using Eq. (7.2). Atoms that emit radiation as a result of electronic decay from an excited state do not change mass significantly; all the energy released comes from the excess kinetic energy of the electron.

Radiation travels at the speed of light c. If you imagine a wave passing a point at this velocity (Fig. 7.1), you can see that the frequency ν, velocity c, and wavelength λ (Greek symbol "lambda") are related by

$$\nu = \frac{c}{\lambda}$$

So, high-frequency radiation will have a small (short) wavelength, and low-frequency radiation will have a long wavelength. Returning to

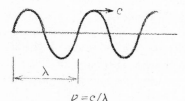

$$\nu = c/\lambda$$

FIG. 7.1 RELATIONSHIP BETWEEN FREQUENCY AND WAVELENGTH.

Eq. (7.2), we see that photons (quanta) of high-frequency radiation will be highly energetic. The energy per photon increases going from long radio waves to shorter light waves to very short x-rays to very, very short gamma rays. X-rays, for example, typically have energies of the order of a few tens of kev.

Just as the electrons can exist in excited states, so can the nuclei. However, the energy required to raise a nucleus to an excited state is much greater than for electronic excitation. Energy jumps of the order of 1 Mev are typical for lightweight nuclei. The excited nuclear states are very unstable, and the nucleus tends to return to the stable "ground" state in a small fraction of a second by emitting "hard" (high-energy or "gamma") photon radiation. The number of neutrons and protons in the nucleus is unchanged, and so the isotope is not changed by the gamma emission.

There are many nuclear reactions that *do* change the basic nature of the atom, and these constitute what is know as "radioactivity." Many nuclei, even some that occur "naturally" on earth, are unstable and change to some stable form by emission of some sort of particles. There are several different modes of radioactive decay, and these are usually grouped according to the nature of the emitted particles. In "alpha emission" the "particle" emitted consists of a cluster of two neutrons and two protons; this is just the nucleus of a helium atom, and the alpha particle is nothing more than a helium nucleus. This mode of decay is common for the heavy elements where the binding energies are lower, and is frequently accompanied by gamma emission. Since the alpha particle carries off protons and neutrons, the basic nature of the atom is changed. An important example is the decay of plutonium,

$$_{94}Pu^{239} \rightarrow \,_2He^4 + \,_{92}U^{235}$$

In writing equations describing nuclear reactions, it is customary to use the notation employed above. The fore-subscript (e.g., 94 for Pu) is the charge of the nucleus, i.e., the number of protons in the nucleus. The superscript (e.g., 239 for Pu) is the "mass number," i.e., the sum of the number of neutrons and protons in the nucleus. This notation makes balancing of nuclear particles very easy, since $94 = 2 + 92$ and $239 = 4 + 235$. Moreover, anyone familiar with the basic isotopes can figure out right away what new elements will result from alpha emission of any element.

Another important type of radioactivity involves emission of "beta" (β) particles from the nucleus. The beta particle is simply an electron with positive charge ("positive beta") or negative charge ("negative beta"). Beta emission provides a mechanism for changes to occur in the charge of a nucleus without a change in mass. The result of a negative beta emission is that a neutron is effectively converted into a proton; scientists would describe this by

$$_0n^1 \rightarrow \ _{-1}e^0 + \ _1H^1$$
$$\text{neutron} \quad \text{beta} \quad \text{proton}$$

Note that the subscript and superscripts both "balance" on the two sides of this nuclear-reaction equation. The positive beta particle is called a "positron," and the designated by $_1e^0$. As an example of positron emission,

$$_7N^{13} \rightarrow \ _1e^0 + \ _6C^{13}$$

describes the decay of an unstable isotope of nitrogen to a stable form of carbon.

These are just some of the important nuclear reactions that occur spontaneously. All these occur at the whim of the nucleus, some much more rapidly than others. Statistical studies of all sorts of radioactive decays indicate that the number of decays that occur in a small time interval is proportional to the number of radioactive atoms that are present at that time. Thus, if Δt denotes time interval and $-\Delta n$ the decrease in the number of radioactive atoms over this time interval, the "radioactive-decay law" is

$$\frac{-\Delta n}{\Delta t} = an$$

where a is a constant that depends upon the decay reaction. If you know a little calculus, you will be able to solve this equation for the limiting case of very small time changes, and thereby obtain an expression for the number of radioactive atoms present as a function of time,

$$n = n_0 e^{-at} \tag{7.3}$$

Here n_0 is the number present initially when $t = 0$, and e is the mathematician's special number 2.71828. Figure 7.2 shows the form of this "exponential decay law" that applies to all radioactivity. The time scale for the decay could be quantified by giving the value for the constant a. However, it is common practice instead to deal with the "half-life" of the decay, defined as the time required for half the atoms to decay from their

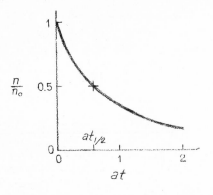

FIG. 7.2   RADIOACTIVE-DECAY LAW.

radioactive or excited state. The half-life is designated as $t_{1/2}$ in Fig. 7.2. Half-lives vary from fractions of a second to billions of years. For example, the much-discussed product of nuclear weapons, Sr^{90} (Strontium 90), has a half-life of about 20 years. So, strontium 90 deposited into the earth's atmosphere as a result of a nuclear war would be only half gone after 20 years, $\frac{3}{4}$ gone after 40 years, $\frac{7}{8}$ gone after 60 years, and all this time would be emitting high-energy radiation as atom-by-atom it decays. The longer the half-life, the longer radioactive atoms are around, and the more slowly they decay. Radiation damage is dependent upon the *intensity* of radiation, and a slowly decaying isotope will emit relatively slowly. So, naturally occurring U^{238}, which has a half-life of 4.5×10^9 years, decays so slowly that its radiation is not at all harmful by comparison with other radioactive materials around.

The earth has been around long enough that the elements with short half-lives have decayed away, leaving only those with long half-lives and those which are "stable." The nuclear arrangements for stable nuclei are shown in Fig. 7.3. It is remarkable that these form a narrow band, with the light elements having about the same number of neutrons as protons, and the heavier elements having progressively more neutrons than protons.

In addition to spontaneous (radioactive) nuclear reactions there are many reactions that occur when nuclear particles collide. For example, in a nuclear reactor neutrons, betas, and gammas are flying around every which way, continually colliding and reacting with the nuclei of stationary atoms of a variety of elements. The details of the reaction depend upon the energy of the approaching particle and the directness with which it strikes the nucleus. The energy of the approaching particle is easy to quantify, but the directness of the impact is not, and so one thinks of "probabilities" of certain things happening as a result of the collision; this statistical approach avoids the problem of dealing with the impact more precisely. The most important nuclear reactions are listed thus:

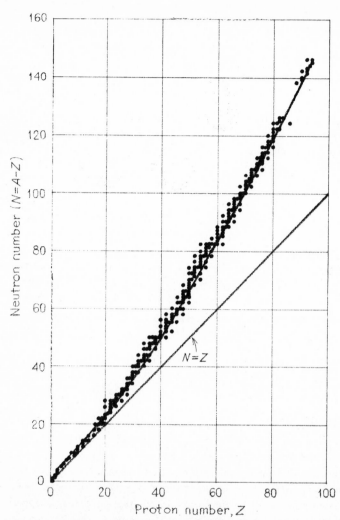

FIG. 7.3 NEUTRON-PROTON PLOT OF STABLE NUCLIDES (INCLUDES RADIO-NUCLIDES WITH HALF-LIVES GREATER THAN 1,000 YEARS).

Scattering:

Neutron hits nucleus and glances off at a reduced speed; nucleus is set in motion, perhaps excited, but otherwise unaffected.

Absorption:

Neutron (usually slow) is captured by the nucleus, which then "decays" by emission.

Fission:

Neutron "splits" the nucleus into two light nuclei, forming two lighter atoms (often radioactive).

Fusion:

Deuteron particle (proton plus neutron) is captured and a proton or neutron is emitted.

Figure 7.4 depicts examples of these four important types of nuclear reactions. There are others that occur in nuclear reactors and fusion power experiments, and you can read about these on your own.

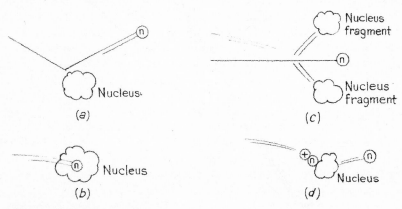

FIG. 7.4 SCHEMATICS OF NUCLEAR REACTIONS. (*a*) SCATTERING, (*b*) ABSORPTION, (*c*) FISSION, (*d*) FUSION.

The various possibilities for nuclear reaction are usually expressed in terms of the "cross sections" for the various events. This arises from experiments in nuclear physics where the approaching particle is assumed to have negligible size and the probability for a particular reaction is treated as being proportional to the projected area of the target nucleus. Figure 7.5 shows a simple model used in the conception of the cross section. A beam of particles with intensity I_0 (number per second per square centimeter) impinges upon a target with thickness t. Some of the approaching particles collide with target atoms; others do not. The ones that come through unscathed make up a weaker beam with intensity I. Experiments (and theory of a more sophisticated nature) show that the emerging and incident beams may be related by

$$I = I_0 e^{-\sigma N t} \tag{7.4}$$

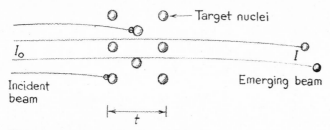

FIG. 7.5 THE "CROSS SECTION" DETERMINES THE PROBABILITY OF COLLISION OF PARTICLES IN THE INCIDENT BEAM.

where N is the number of target atoms per cubic centimeter, and σ (the lowercase Greek letter "sigma") is the "cross section" for the particular sort of collision being studied. The product σNt is dimensionless; σ has the dimensions of area, usually expressed in cm^2. Since large nuclei have diameters of the order of 10^{-12} cm, it is not surprising that σ is typically of the order of 10^{-24} cm^2. It has become commonplace to define 1 "barn" as 10^{-24} cm^2, and now cross sections are normally stated in barns.

When a neutron beam strikes a target, several things happen. There will be some fissions, there will be some scattering, there will be some absorption, and some neutrons will get through without collision. Each of these reactions has its own cross section, and the intensity of the beam that emerges without collisions of any sort is again given by Eq. (7.4), with σ replaced by the *total* cross section, i.e., the sum of the cross sections for the various possible reactions,

$$\sigma_{\text{tot}} = \sigma_{\text{fis}} + \sigma_{\text{abs}} + \sigma_{\text{scat}} \qquad (7.5)$$

The probability of each type of reaction upon collision is given by the ratio of the cross section for that reaction to the total cross section, e.g.,

$$p_{\text{fis}} = \frac{\sigma_{\text{fis}}}{\sigma_{\text{tot}}} \qquad (7.6)$$

The values of the cross sections depend upon the nature and energy of the incident particle. Figure 7.6 shows the total cross section for natural uranium, as measured experimentally. The high peaks are due to high-absorption cross sections for neutrons of particular energies by the isotope U^{238}. We will refer to this curve later to explain why U^{238} is removed from natural uranium in the preparation of nuclear-reactor fuels.

The product of a fission reaction depends upon the manner in which the neutron strikes the nucleus. Again a statistical approach is used to

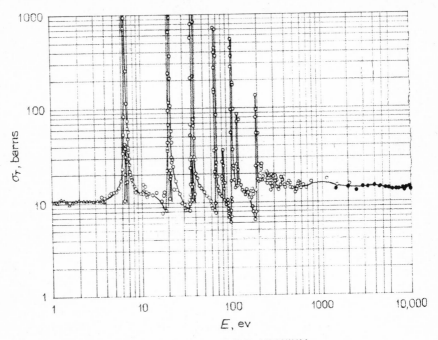

FIG. 7.6 TOTAL CROSS SECTION OF NATURAL URANIUM.

avoid having to deal with this precisely. Figure 7.7 shows the fission-product yield from the fission of U^{235} by slow ("thermal") neutrons. The curve shows the percentage of each sort of nuclide that is yielded in a large number of fission reactions. The fission tends to produce one particle with a mass number around 140 and another with a mass number around 95 ($140 + 95 = 235$ as required to balance the nuclear reaction), and only very infrequently are particles with significantly different mass numbers produced by the fission reaction. Curves such as this have been measured for many fissionable isotopes, and they generally take the same shape.

When a fission reaction occurs, the nucleus is split into two fragments of lower mass number. These nuclides will have a greater neutron-proton ratio than the stable nuclei (see Fig. 7.3), and hence will be radioactive. So, they decay by emission of excess neutrons and other radioactive-decay reactions. Most of the neutrons are emitted within 10^{-12} sec after fission; these are called the "prompt" neutrons. Less than 1% are "delayed" in their emission, but these are still very important in nuclear reactors. The number of neutrons emitted depends on the particular "fission fragments" that happen to split out of the original atom. On the average, U^{235} fissions produce about 2.5 neutrons per

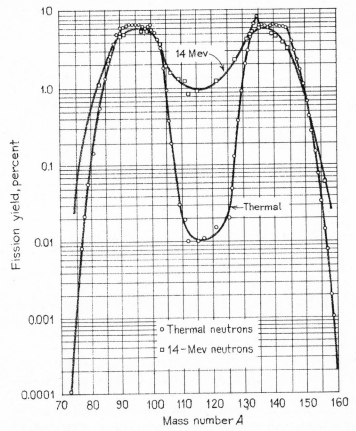

FIG. 7.7 FISSION-PRODUCT YIELD FROM U^{235} BY 14 MEV AND THERMAL NEUTRONS.

fission. It is this excess of output neutrons over input neutrons that permits the nuclear reactor to stay "self-supporting." The energy of the prompt-fission neutrons varies markedly, depending upon the particular decays involved. Most of the neutrons are emitted with energies of the order of 1 to 5 Mev. The delayed neutrons tend to be less energetic, and this is very important in the control of nuclear reactors, as we shall see.

The fission reaction releases energy in a variety of ways. The total energy emission per fission of a U^{235} atom is about 200 Mev. Most of this (167 Mev) is in the form of kinetic energy of the fission fragments, the rest being largely radioactive-decay energy and kinetic energy of the neutrons. The kinetic energy of the fission fragments is ultimately transferred to other atoms by collisions, with the result that the whole thing gets hot. In a bomb things just keep getting hotter and hotter, while in a nuclear reactor a coolant is provided to maintain the fuel at a fixed operating temperature. Measurements of the thermal energy produced per fission of U^{235} indicate about 175 Mev, suggesting that about 8 Mev is produced by collisions of particles other than the primary-fission fragments.

We have outlined the basic aspects of nuclear reactions that we need to develop some understanding of nuclear energy, and will now turn to some of the practical aspects of nuclear power technology.

NUCLEAR POWER REACTORS

In the previous section we described the process of nuclear fission, in which a neutron striking the nucleus of a fissionable atom breaks the nucleus into two lighter pieces that subsequently decay radioactively by throwing off some neutrons. Some of these neutrons will be "captured" by atoms that they strike; others will cause new fissions of atoms. The reaction is said to be "critical" when there is exactly one new fission for each old one; the reaction is then a "chain" reaction. In a critical condition the fission rate and hence energy release per unit time is constant, and thus a steady-state source of energy is provided. The fission products collide with stationary atoms, giving up their kinetic energy, and this increases the temperature of the fuel material. So, if a coolant is provided, energy can be transferred continually to the coolant as heat, and this energy then can be used in a fluid power system of the sort discussed in Chap. 4. The power output from a nuclear reactor can be controlled by controlling the number of neutrons flying about within. This controls the number of fission reactions, and hence the "reactor thermal power." This is the basic idea of nuclear power as it is known today.

In order to achieve criticality, enough neutrons must be produced by each old fission to produce at least one new fission. We say "at least" because in the process of bringing the reactor up to operating power there must be an excess of new fissions so that the power level can rise continuously. If the excess is too great, the power will increase too rapidly, and the reactor becomes a nuclear bomb. Reactors are of course designed such that it is impossible for the power-rise rates used in bombs to occur, but start-up still is a phase of operation that must be done very carefully by a trained operator. As with airplanes, the "take-off" is the most dangerous moment in the operation of a nuclear reactor.

The heavy atoms (those with high mass number) are best suited for fission reactions because their fission fragments will have a relatively large excess of neutrons to release upon radioactive decay. The isotopes U^{235}, U^{233}, Pu^{239}, and Pu^{241} (Pu stands for plutonium) are commonly used because they have long half-lives and produce a number of neutrons upon fission. The uranium isotopes are abundant naturally, and the plutonium isotopes can be created from uranium in a "breeder" reactor. It is impossible to establish a chain reaction in natural uranium because the U^{238} captures too many of the neutrons that otherwise could cause fissions of U^{235} (see Fig. 7.6). Figure 7.8 shows the fission cross section for U^{235} at relatively low neutron energies. Note that the fission cross section is higher at low energies; consequently it is much easier to fission U^{235} with "slow" neutrons than "fast" ones (the neutron is around the nucleus longer and has a better chance to score). These considerations suggest that one should try to establish a critical chain reaction by removing some of the U^{238} to eliminate its undesirable neutron capture, and then somehow slow down the neutrons so that they will have a high probability of causing fissions when they strike U^{235} nuclei. This was indeed the approach used in the early days of reactor development. We will mention some newer concepts shortly.

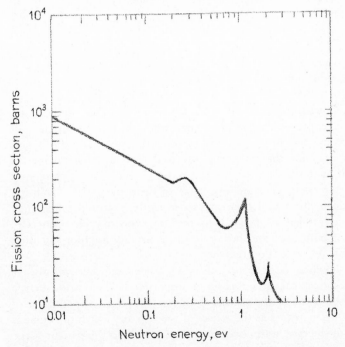

FIG. 7.8 FISSION CROSS SECTIONS FOR U^{235} AT LOW ENERGIES.

How can one slow down neutrons without losing them by capture? One needs a material with a high scattering cross section and a low capture cross section; the scattering results in transfer of kinetic energy from the neutrons to the scattering atoms, and this slows down the neutrons and agitates the scattering atoms. Eventually an equilibrium condition is obtained in which some neutrons gain energy and others lose energy when they strike the agitated scattering atoms. In this condition there is no longer any net transfer of energy between the neutrons and the atoms, and they are "in thermal equilibrium." So, neutrons that have been slowed to this level are called "thermal" neutrons. It is the thermal neutrons that one wants to use to cause U^{235} fissions. Can you guess which atoms are best for the job of slowing down or "moderating" neutrons? If you have ever played marbles, you know that a small marble keeps most of its energy when colliding with a large one, and is most effective in giving its energy to another marble when the two marbles are about the same size. So, a good moderator atom should have about the same mass as a neutron. Therefore, hydrogen, the lightest atom, makes the best moderator, and water has been used as a moderator in many reactors. "Heavy water," which has the hydrogen isotope $_1H^2$ instead of the more abundant form $_1H^1$, has also been used because of its lower capture cross section. Water and graphite are the primary moderators in use today.

In order to make efficient use of the neutrons that are available in a reactor, they must not be allowed to escape. So, a nuclear reactor is surrounded by a "shield" that does two things. First, it is designed to reflect as many neutrons as possible back into the reactor core; and second, it is designed to capture those neutrons that cannot be reflected to prevent them from coming out into the environment. A thin layer of heavy atoms is sometimes used for the reflecting job, and then a thick blanket of hydrogen-rich material is used for the outer shield. The hydrogen is good because escaping neutrons lose most of their energy in a collision with a hydrogen atom and then are more easily captured when they collide with a heavier atom, such as lead or even oxygen. Concrete, which contains both light and heavy atoms, is usually used for shielding in large central power station reactors.

Since no shielding is perfectly reflecting, some neutrons will always be lost. The neutron leakage increases in proportion to the reactor surface area, i.e., to the *square* of the reactor diameter. However, the number of fissionable atoms increases as the *volume* of the reactor, which of course is proportional to the *cube* of the diameter. So, by making the reactor large enough, a "critical" situation can be obtained. Reactors, and bombs, have "critical masses" below which they just will not go critical. Indeed, one of the safety features of power reactors is that they are built with a limited amount of fissionable material so that they are only barely able to go critical, quite unlike a bomb.

Now imagine that you are a nuclear-reactor engineer. You have some fuel, understand the need for moderation, and have solved your shielding problem. How are you going to control the number of fission reactions and hence vary the reactor power during operation? . . . (the dots represent your thinking about it for a few minutes). One way would be to change the amount of fuel in the reactor core; but the fuel will be hot and hard to handle, and the AEC (Atomic Energy Commission) safety guys aren't about to let you control your reactor this way. Another way would be to vary the amount of moderator present; for example, you could move some graphite in and out of the core, thereby controlling the number of thermal neutrons flying about. The trouble with this is that you reduce the reactor power by increasing the number of fast neutrons, and in an emergency you could very likely be increasing the radiation hazard when you are trying to shut down the reactor. The AEC would not be very fond of this either . . . (back to the drawing boards). How about inserting or withdrawing a material that loves to capture neutrons? Now to increase the power, you gingerly withdraw your "control rod," thereby allowing the number of neutrons to build up slowly, and then you can push the rod back in a ways to hold the neutron population at the level required for the desired reactor power. At last . . . you have invented a safe scheme for reactor control! And if you make things "fail-safe," so that if anything goes wrong the control rods fall into the core by gravity, thereby shutting things off automatically, and do a good job in your engineering, the AEC just might grant you an operation permit and supply you with nuclear fuel.

Now you should have a fairly complete picture of the basic features of a nuclear power reactor; fissionable material, appropriately enriched with U^{235} by removal of U^{238}; a moderator to produce slow neutrons; shielding to minimize neutron escape and protect the public; and a "poison" type of control system designed to be supersafe. You allow the neutron population in the reactor to build up, and then hold it at the desired operating level. The energy released by the fission is converted to internal energy in the reactor core, and you take this energy out with the reactor coolant. The hot fluid is then worked over in a fluid energy-conversion system designed with due regard for public safety.

As the nuclear reaction goes on, the amount of fissionable material in the core gradually decreases. The control rods gradually are moved out of the core to maintain the reactor power as this process of "burnup" goes on. Eventually the fuel will be depleted to the point where the fuel elements must be removed and replaced with new elements containing a higher percentage of fissionable material. The old plates are then processed to remove as much fissionable material as possible, and then disposed of in some manner that is appropriate to their degree of radioactivity. Commercial power reactors are shut down for partial refueling and maintenance about once each year. The replacement of the entire reactor core thus extends over a period of several years, and in reality is always going on, much like the painting operation on the Golden Gate Bridge.

Several different types of coolants have been used in power reactors. The "boiling-water reactor" (BWR) has become the workhorse of the commercial power industry. Recalling that water is a good moderator, you can see that if the boiling produces too much steam, the number of moderating hydrogen atoms in the core will go down (steam is less dense than liquid water, remember); this in turn will reduce the number of slow neutrons in the core, and hence will reduce the fission rate and reactor power. When combined with poison-type control rods, good and safe control is provided. Typically the water flow is only partially vaporized in the reactor, and the liquid is recirculated back through the reactor after the steam has been removed. The steam can, in principle, be fed directly to a steam turbine. But unfortunately this sort of power plant does not have good load-following capabilities; as the turbine load increases, the turbine starts to slow down, and the turbine speed control then asks the reactor for more steam by opening a valve between the reactor and turbine. But this tends to drop the reactor pressure, allowing steam within the reactor to expand, thereby increasing the "void space" in the water moderator and thus reducing the reactor power. The control rods must work hard to overcome the resulting loss of moderator. To get around this problem, boiling-water reactors often bypass a portion of the flow around the turbine, "dissipating" the available energy in the bypass steam by pressure drop through a valve. Then, when the turbine asks for more steam, the bypass flow is simply diverted to the turbine, and the reactor continues to operate as if nothing had happened. But this scheme is very wasteful of available energy, and so in practice is used only for small "load trimming." A still better way to control the plant output is to change the recirculating-water flow rate, and this approach is becoming more common in new nuclear power systems.

The fuel elements in BWRs are typically tubes of corrosion-resistant metal filled with enriched uranium oxide. The tubes are arranged in a "fuel bundle" through which the water flows vertically upward along the tubes. Spacers are used to maintain proper spacing between the fuel rods and thus to prevent the formation of undesirable "hot spots" which could

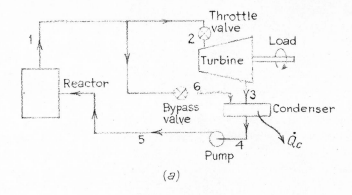

(a)

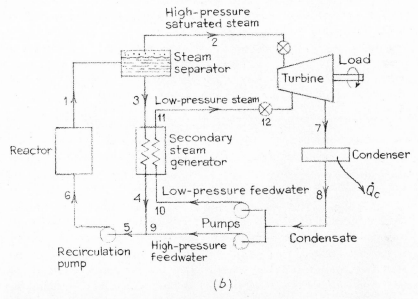

(b)

FIG. 7.9 BOILING-WATER-REACTOR
SYSTEMS. (a) DIRECT SYSTEM,
(b) DUAL CYCLE SYSTEM.

lead to premature failure of the fuel rod. The rod bundles form "sub-
assemblies" that are inserted into the core in a matrix style. The neutron
population in the center of the reactor core is higher, and hence the fuel
bundles in the center of the core normally are loaded with less fissionable
material (lower enrichment) in an attempt to provide uniform power
output across the reactor. Figure 7.9 shows some alternative plumbing
schemes for BWR reactors. Figure 7.10 shows some details of a commercial
BWR.

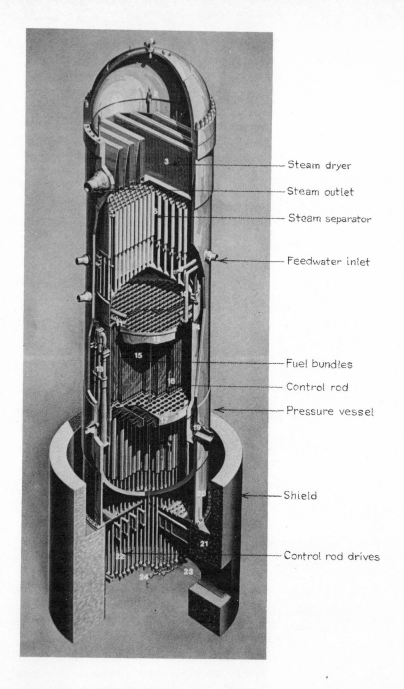

— Steam dryer

— Steam outlet

— Steam separator

— Feedwater inlet

— Fuel bundles

— Control rod

— Pressure vessel

— Shield

— Control rod drives

FIG. 7.10 THE BWR/6 SYSTEM. COURTESY OF THE GENERAL ELECTRIC COMPANY.

Another type of water-cooled reactor is the "pressurized-water reactor" (PWR), in which the water is maintained at a high pressure to prevent bulk boiling and the accompanying formation of large-scale vapor "voids." The reactor pressure is typically 2,000 psia, and the liquid water leaves the reactor at temperatures around 560°F. This hot water is used to boil steam in a secondary circulation system, for use in a steam turbine. Alternatively, the water flow from the reactor can be passed through a valve, which drops the pressure and "flashes" some of the liquid to vapor. The PWR, which was developed initially as a concept in the U.S. Naval Submarine Reactor Program, has now become one of the important reactor types in the commercial nuclear power industry. Figure 7.11 shows some typical PWR plumbing arrangements.

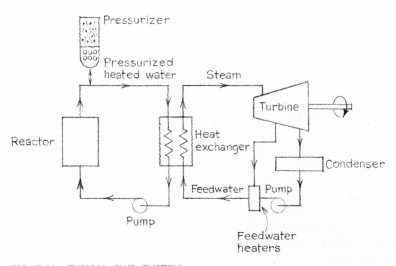

FIG. 7.11 TYPICAL PWR SYSTEM.

The load-following capability of a PWR derives from the expansion and contraction of the water with temperature. As the turbine load increases, valves are actuated, with the effect that the reactor inlet water becomes colder and more dense. This increases its ability to moderate the fast neutrons, and hence the slow-neutron population increases, increasing the reactor power. The control rods are then used mainly to compensate for long-term changes in the neutron density due to "fuel burnup" and "fission-product poisoning." The converse effect is that an increase in temperature expands the water, reducing its ability to moderate fast neutrons and hence reducing the population of slow neutrons and therapy reducing the reactor power. Thus, the PWR has a large "negative temperature coefficient," meaning that a large negative change in slow-neutron population is produced by positive changes in temperature. This has obvious safety features which are fully capitalized upon in the engineering of such reactors. For example, the Shippingport PWR once had a coolant-flow failure during which the "scram rods" (control rods that are shot into the reactor to reduce the slow-neutron population in case of emergency) were not activated. The reactor showed only a slight increase of pressure and a slow drift upward in temperature that was easily corrected by insertion of the main control rods.

A disadvantage of the PWR and BWR systems is that the working fluid is water. This presents corrosion problems, and in addition limits the maximum working-fluid temperature at tolerable pressures. Another concept, the "gas-cooled reactor" (GCR), gets around these problems by using a gas such as carbon dioxide, helium, or nitrogen as the working fluid. The gas heated in the reactor may be used directly in the power-conversion cycle, or else used to heat the working-cycle gas. Figure 7.12 shows these possibilities schematically. In either case the reactor basically plays the role of the combustion chamber in the Brayton cycle (see Chap. 4). Alternatively, the hot-reactor outflow could be used to generate steam in a secondary heat exchanger, which then replaces the boiler in a conventional steam power plant.

In a GCR graphite is used as the moderator. The fuel may be mixed with the moderator, but more often they are kept separate for metallurgical and servicing reasons. The fuel elements are typically bundles of tubes filled with slightly enriched uranium oxide. Since the convective heat-transfer coefficient for gas flow is typically considerably less than for liquid flow (the molecules are further apart and thus less able to soak up energy from the heated surfaces), large temperature differences between the fuel elements and the gaseous coolant are typical in GCRs. In some designs this is reduced by the addition of heat-conducting fins, in the spirit of the fins on automobile radiators.

Other fluids that have been used in experimental and commercial nuclear reactors include liquid metals such as sodium or an organic liquid such as terphenyl. The advantage of the organic liquids is a low vapor pressure at a high temperature. The advantage of liquid metals is the high convective heat-transfer coefficients that they afford. In these reactor systems the coolant is used to heat the cycle working fluid (usually steam) in a closed power cycle. The liquid-metal reactors require a separate moderator such as graphite. Since organic fluids contain hydrocarbon molecules, their hydrogen atoms can do the moderating in organic-cooled reactors.

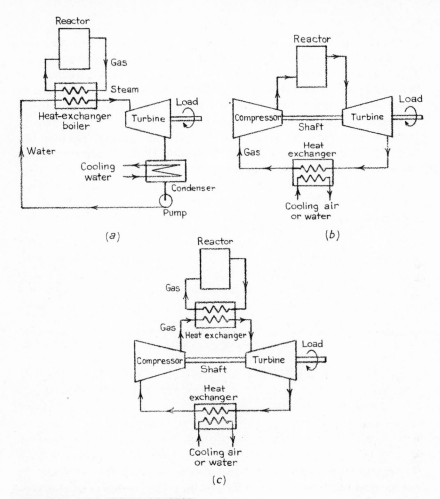

FIG. 7.12 GAS-COOLED-REACTOR SCHEMATICS. (*a*) THE INDIRECT CLOSED CYCLE, GAS TO WATER, (*b*) THE DIRECT CLOSED CYCLE, (*c*) THE INDIRECT CLOSED CYCLE, GAS TO GAS.

239

Other experimental "liquid-fueled reactors" (LFR) in which the nuclear fuel is actually carried with a liquid coolant have been built. In these systems the fuel is carried in a liquid as small particles or is actually dissolved into the carrier coolant. The reactor is simply a large bulge in the coolant-circulation loop where enough fissionable material is concentrated to produce a critical mass. There are many technological problems that remain to be solved with LFRs, and it is not clear that they will ever compete with the BWR, PWR, and GCR as mainstays of the commercial nuclear power industry.

BREEDER REACTORS

Earlier we mentioned that some isotopes of uranium are more easily fissioned than others, and that U^{238}, the most abundant naturally occurring form, does not fission at all (it is "fertile"). However, one of the reactions that can occur in a reactor is the nonfission capture of a neutron by a U^{238} nucleus, which produces a series of reactions culminating in the conversion of the nucleus to a fissionable form of plutonium, Pu^{239}.

$$_0n^1 + {}_{92}U^{238} \rightarrow {}_{92}U^{239} + \gamma$$

$$_{92}U^{239} \rightarrow {}_{93}Np^{239} + {}_{-1}e^0$$

$$_{93}Np^{239} \rightarrow {}_{94}Pu^{239} + {}_{-1}e^0$$

The capture first produces U^{239} and gamma radiation. The U^{239} decays radioactively with a half-life of 24 min to a radioactive isotope of neptunium, which in turn decays with a half-life of 2.3 days to the fissionable plutonium. The two decay reactions release negative beta particles. This process is called "conversion"; it produces a long-lived atom of fissionable material from an atom of fertile material. There are other conversion reactions, but this is perhaps the most important one.

Some fissionable plutonium is produced by the chance reactions described above in any nuclear reactor. In some cases this is retrieved from spent fuel and then used in other reactors. In fact, "converter" reactors have been built just for the purpose of plutonium production. It is even possible to convert more atoms than are consumed in the fission reactions that produce the neutrons. A reactor in which more fissionable nuclei are produced than consumed is called a "breeder" reactor. The breeder reactor literally produces more atoms of fissionable material than are used in the fission reactions! Breeder reactors therefore provide a way to make more complete use of uranium ore, and obviously are going to have an enormous impact upon the nuclear power industry.

The breeding reactions given above are most effective when the reactor is populated with very energetic ("fast") neutrons. In a fast breeder reactor the chain reaction is maintained with fast neutrons having an average energy of around 1 Mev by fission of U^{235} and Pu^{239}. These fissions produce sufficient neutrons to maintain the fission reaction and enough extra neutrons to convert U^{238} to additional Pu^{239}. Slow breeder reactors, which use thermal neutrons to convert a fertile isotope of thorium to fissionable U^{233}, are also possible, and both types of breeders are now operating experimentally. However, the major breeder-reactor development effort is being concentrated on fast breeder reactors.

Fast breeder reactors do not need to have many slow neutrons around, and so they do not use a moderator. This means that water cannot be used as the coolant, and instead gases or liquid metals must be used. A breeder reactor will have an additional component, the "blanket," which may surround the reactor core between the fuel and the shielding or may be spread throughout the reactor. The material that is to be converted is placed in this blanket and is retrieved after the conversion has taken place. The rate of breeding is measured by the "doubling time" of the reactor, the time required for conversion of twice as many fertile atoms to fissionable atoms as have been consumed by fissions during the breeding process. At the end of one doubling period a breeder reactor has produced enough fissionable material to refuel itself and in addition fuel another reactor! It appears that 7- to 10-year doubling periods will be typical for efficient commercial breeder reactors.

Slow reactors have the desirable safety feature that, because of the negative temperature coefficient, they are self-regulating; that is, they are designed to self-compensate immediately for unintentional increase in reactor power. However, in a fast reactor the expansion of the coolant as a result of overheating can lead to an increase in reactor power, which in turn increases the overheating. This unstable condition is fortunately offset by the fact that the jittering motion of hot atoms in the fuel plates affects the ability of their nuclei to capture fast neutrons. This is viewed as a "Doppler" effect, since it relates to the relative speeds of the nuclei and neutrons. The net effect is that the U^{238} capture rate increases with increased temperature, and this helps *reduce* the number of fast neutrons as the temperature increases. The reactor can be designed so that the Doppler mechanism overrides the effect of coolant expansion, with the net result that the reactor power will *decrease* as the temperature increases, as desired for safe and stable operation.

At the present time there is considerable interest in helium-cooled fast breeder reactors, and at least one commercial reactor of this type is now under construction. The hot helium is used to boil steam in a separate heat exchanger, and the steam is then used in a conventional power cycle. The use of gaseous helium coolant is motivated by its low rate of neutron moderation and capture in comparison with liquid-metal coolants, by its inert chemical nature (in contrast to the corrosive nature of liquid metals), and by the fact that the helium does not become radioactive and hence does not pose a safety threat should it leak from the reactor. We are now in the very early stages of nuclear power development. Looking ahead, it seems very clear that some mix of breeder reactors, slow reactors, and converters will be needed to provide the optimum use of nuclear energy resources at a tolerable cost.

RADIOISOTOPE ENERGY SOURCES

The first nuclear energy sources to operate in space were "radioisotope" devices. These devices do not contain a self-sustaining chain reaction but instead utilize the energy released by slow radioactive decay. The particles released by the decay collide with stationary atoms, increasing the temperature of the assembly. Energy is then extracted from the assembly as heat and converted to electricity, usually in a thermoelectric generator of the type discussed in Chap. 6. For example, the SNAP 10A power system had a radioisotope power source with a thermal output of 35 kw and an electrical output of 0.58 kw (you can compute the system efficiency). Heat was removed from the reactor by a liquid-metal coolant known as "NaK," a mixture of sodium and potassium. The NaK carried the energy to the thermoelectric converters, which were arranged in panels along the skin of the space vehicle. Only a small portion of the reactor power was converted to electrical energy, and the rest (the "waste heat") was radiated away from the vehicle by the hot skin. Figure 7.13 shows some details of this system.

A number of other isotopic power sources have been built for space use. The SNAP 27 placed in operation on the moon by Apollo astronauts uses the decay of plutonium 238 and produces around 70 watts of power. The plutonium decays with a half-life of 86 years, and the design life of the power system is 1 year. Other isotopes with shorter half-lives are of interest because they afford higher energy output for a given system weight. For example, curium 242, with a half-life of only 183 days, offers a power density of around 1,000 watts/gram of material, considerably more than the 7 watts/gram available from the slower decay of Pu^{238}. The SNAP 11 system used this curium isotope on a moon probe for which a 90-day system life was quite adequate.

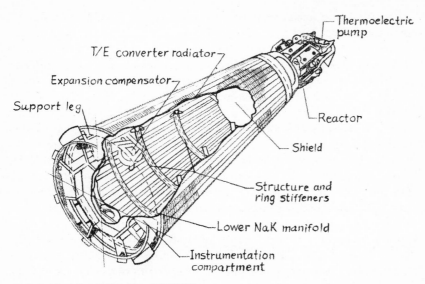

FIG. 7.13 THE SNAP 10A REACTOR.

Devices that convert the nuclear radiation energy directly to electrical energy are also being developed. These are not heat engines and hence are not limited by the Carnot efficiency, although their practical efficiencies are very low. They are capable of producing very small amounts of power for a long time, and will undoubtedly be used for such things as cardiac pacemakers. The "betacel" captures beta particles emitted by isotope decay. These devices operate at around 1 volt. In the "fission electric cell" charged fission fragments move across a vacuum to a collector plate, giving the plate a strong positive charge. Electrons can then be pulled around through an external circuit to the collector plate. The fission fragments have a very large fraction of the energy released by the radioactive decay, and they carry a large number (around 80) of positive charges; so the fission electric cell tends to be a very high voltage device. With further research and development, devices like these may be developed to the point where they have some commercial application.

FUSION POWER

Fission is a nuclear reaction in which a nuclei is split into lighter elements. In contrast, "fusion" is a nuclear reaction in which two light particles unite to form a heavier atom. Fusion reactions provide the primary source of the sun's energy, and the great hope of the future as the ultimate energy source for man.

The particular fusion reactions that can be made to occur in a fusion reactor include

$$D + D \rightarrow T + n + 3.2 \text{ Mev} \tag{7.7a}$$

$$D + D \rightarrow T + p + 4.0 \text{ Mev} \tag{7.7b}$$

$$D + T \rightarrow He^4 + n + 17.6 \text{ Mev} \tag{7.7c}$$

$$D + He^3 \rightarrow He^4 + p + 18.3 \text{ Mev} \tag{7.7d}$$

Here D stands for the deuteron or deuterium nucleus ($_1H^2$), a heavy isotope of hydrogen, T stands for a triton or tritium nucleus ($_1H^3$), a rarer and heavier hydrogen isotope, p is proton, n is neutron, and the energies refer to the kinetic energies of ejected particles. These reactions can be made to occur by slamming the reacting nuclei together at very high velocity. The high-speed impact is needed to overcome the strong repulsive forces between the like charges of the positively charged nuclei. Such impacts will occur naturally if the gas containing the reactants is hot enough. Unfortunately this requires temperatures that are hard to imagine, temperatures of the order of 400 *million* °K! Such a "temperature" cannot be measured with mercury-in-glass thermometers, but instead is defined in terms of the kinetic energy of the particles. The average kinetic energy of a particle in a simple gas is related to the gas temperature by

$$E = \tfrac{3}{2}kT \tag{7.8}$$

where k is the Boltzmann constant, 8.62×10^{-5} ev/°K, and T is the absolute temperature. Equation (7.8) is used to define "temperature" beyond the range where conventional thermometers melt. The fusion temperature of 400×10^6 °K corresponds to a particle energy of

$$\tfrac{3}{2} \times 8.6 \times 10^{-5} \times 4 \times 10^8 \text{ ev} = 52 \text{ kev}$$

Since the ejected particles have energies considerably in excess of 52 kev, they should be able to kick two more reactant nuclei up to this energy and thereby sustain the fusion reaction with lots of energy to spare.

Atoms begin to shed their electrons at temperatures of a few thousand degrees Kelvin. At one million degrees Kelvin the electrons are practically all stripped off, and the positively charged nuclei and free-floating electrons form a "plasma." The presence of free electrons makes it easy for the plasma to conduct electricity, and this fact is very important in the control of fusion reactions. A hot plasma will tend to lose energy by thermal radiation, and if it loses more than it gains from the fusion reactions, the reaction will shut off, much as a match blows out. Each fusion reaction has its own "ignition" temperature that must be attained if the reaction is to be self-sustaining. The deuterium-tritium reaction [Eq. (7.7c)] ignites at around 45 million °K, while the deuterium-deuterium reaction requires 400 million °K. Since the problem of making a fusion reactor work is largely that of keeping the plasma hot, and this is easier to do at 45 million °K than at 400 million °K, it is probable that the D-T reaction will be employed in the first successful fusion-reactor power plants.

In order to contain these ultrahot plasmas, the electrical properties of the plasma are used. The idea is to apply a magnetic field to the plasma that "pinches" the plasma and thereby confines it to the center of the magnetic field, away from the vessel walls. Figure 7.14 shows a typical sort of toroidal geometry in which the plasma current flows around and around the torus (remember the right-hand rule in Chap. 6?). It tells us that the force exerted by the magnetic field on the current is toward the center of the torus. Thus, the magnetic field tends to pinch the plasma into a looplike ring in the middle of the magnet. This pinch compresses the plasma and thereby increases its temperature. Magnetic constriction of this sort is the primary means for achieving the high plasma temperatures needed to initiate a controlled fusion reaction.

Unfortunately the system of Fig. 7.14 does not work exactly as advertised above. The plasma exhibits instabilities that cause it to wander and deform within the torus, eventually striking the cold walls and quenching. It appears that these plasma-instability problems can be solved by adding stabilizing windings, distorting the torus, or some

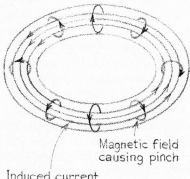

Magnetic field
causing pinch

Induced current
in plasma

FIG. 7.14 IDEA OF MAGNETIC CONTAINMENT. PINCH EFFECT IN A TORUS.

combination of the two. The "stellerator" machines in the United States and the Tokamak machines in Russia are of this type, and the Tokamak devices were the first to reach temperatures a mere order of magnitude below those needed to sustain the fusion reaction. At this writing no fusion reactor has operated with a sustained fusion reaction, but some experts believe that Tokamak or stellarator devices will do so before 1985. Other experts place their bets on devices that use high-energy lasers to heat the plasma to fusion temperatures.

Once a continuous (or perhaps pulsating) fusion reaction is obtained, the problem of converting the kinetic energy of the ejected particles to electrical energy remains. Most of the energy of the deuterium-tritium reaction is in the kinetic energy of the ejected neutron. A good many of these neutrons will knife their way through the magnetic-field coils and escape from the plasma chamber. Once outside they can be moderated in an appropriate fluid, thereby heating the fluid. This hot fluid then can be used to boil steam, and from there on out we have a conventional power generation system. Liquid metals, especially lithium, are attractive for this purpose. High convective heat-transfer coefficients (and hence high power densities) can be obtained with liquid-metal coolants; they do not capture neutrons appreciably. A problem is that liquid metals also carry currents, and the power required to pump them through pipes in a magnetic field can be more than without the magnetic field. Lithium has the advantage that tritium, a scarce isotope of hydrogen that is one of the fusion reactor fuels, can be bred from lithium. So, nonconducting salts containing lithium atoms ($LiF-BeF_2$, or "flibe") are also of some interest. Research into the thermal and physical properties of these fluids is now under way in parallel with the research and development of fusion reactors, but the construction of the world's first power-producing fusion reactor is still many years away.

An alternative concept for fusion power requires an "open" reactor into which the fuel is squirted and from which it flows continuously or in pulses. Magnetic containment and constriction and laser-heating concepts have been explored but are far from the point of providing a sustained thermonuclear reaction. If such a device could be obtained, the electrical power might be generated by passing the output plasma through a "magnetohydrodynamic" generator (see Chap. 6). This may well be the ultimate means for large-scale fusion power generation.

Elsewhere we have mentioned the fact that fossil-fuel reserves are dwindling, and that nuclear reserves for fission reactors are finite. But the energy reserves for fusion power are virtually infinite. If the deuterium-deuterium reaction can be exploited, the limiting factor will be the amount of water in the oceans; one hydrogen atom in every 6,700 is deuterium, and so there is practically an inexhaustible supply of deuterium. One estimate is that the energy available from deuterium in the oceans is of the order of 10^8 times as large as the world's initial fossil-fuel energy reserves, and of the order of 10^7 times the world's fission energy reserves! On the other hand, if the deuterium-deuterium reaction cannot be made to go but the deuterium-tritium reaction can, the fusion energy will be limited by the tritium supply, most of which would have to be created by breeding from lithium. Estimates of the lithium reserves suggest that the tritium supply would limit fusion energy to the order of magnitude of the initial fossil-fuel reserves. Thus, we might expect lithium to be in short supply after a hundred years of deuterium-tritium fission power, and the scientists and engineers of that era will be working hard to develop a deuterium-deuterium reactor.

Your ability to deal with nuclear energy technology in a sensible way will require some basic background in radioactivity as it relates to public health. We'll take this up in the next section.

RADIATION AND PUBLIC SAFETY

There are several important quantitative measures of radioactivity. The "curie" is defined as that amount of any radioisotope that undergoes 3.7×10^{10} disintegrations per second. The mass that makes up a curie depends upon the half-life of the radioactive isotope. For long-lived isotopes such as U^{238}, several tons are required to make one curie, while for a short-lived isotope such as I^{131} a curie would be about the size of a speck of dust. The amount of radioactive material deposited into the atmosphere or to be disposed of as reactor waste is usually given in terms of curies or picocuries ("pico" means 10^{-12}).

In dealing with the biological effects of radiation, the "roentgen," "rad," and "rem" are used as measures of radiation exposure. The roentgen is defined in terms of the amount of x-ray or gamma radiation necessary to ionize a certain number of atoms in a cubic centimeter of air at room conductions. The "rad" is defined as the quantity of any radiation which upon absorption in body tissue is accompanied by the emission of 100 ergs/gram of tissue. The "rem" is the amount of radiation that will deposit the same amount of energy into the tissues of an average man as one roentgen of gamma radiation. Radiation levels in the nuclear industry are commonly expressed in rem, or in thousandths of a rem ("millirem" or mrem).

The controversy over nuclear power abounds with various estimates of the number of curies that will escape into the atmosphere as a result of some hypothetical reactor accident and estimates of the resulting exposures in rem. The Atomic Energy Commission has set standards for permissible radiation; these standards are based on limited data and educated judgment, and sometimes are attacked as being too low. The AEC limit for radiation of an individual in the general public is set at 500 mrem/year to the whole body. The AEC limit for an industrial worker is ten times greater, 5,000 mrem/year. The AEC limits irradiation of bone areas to 29,000 mrem/year, and some experts feel that bone irradiation as high as 291,000 mrem/year is "probably" safe.

Radiation comes from many places. Cosmic rays bombard the world continually, providing an average exposure to the general public of about 50 mrem/year per person. Ground radiation from the earth is also significant, ranging from 180 to as high as 1,800 mrem/year. Buildings contain traces of radioactive materials, and a concrete structure can provide levels up to 100 mrem/year. Food and water expose us each to about 25 mrem/year. All these add up* to an average exposure per person of about 140 mrem/year from *natural* causes. In addition, the average exposure from dental x-rays in the United States is around 55 mrem/year.

Nuclear power plants are designed to provide no more than 5 mrem/year to an individual living just outside the plant. This is a very small contribution to the total radiation exposure, and does not cause much widespread concern. Indeed, the output of radiation from a nuclear power plant is less than the output from a comparable fossil-fuel power station! However, a nuclear accident, either by an overheating and meltdown failure of a reactor core or in transportation of reactor fuels or wastes or in the storage of radioactive wastes, could significantly increase the exposure of some and perhaps many individuals, and this is a real cause for concern. However, the nuclear energy industry and the AEC have strict procedures for hazards analysis, reactor operation control, waste disposal, theft protection, etc., and a remarkable record of safety to go with all these precautions.

The opposition to nuclear power development has come from both the technically uneducated and some very knowledgeable scientists. Both types of people also support the rapid development of nuclear power. There are risks to be taken in this venture; there are also risks in not providing an adequate supply of energy to meet the various real needs of our society. Judgments in debates over the development of nuclear power invariably revolve around the issues of the *values* of society, and

---

* If you wonder why the numbers add up to more than 140 mrem/year, it is because a person is not steadily exposed to radiation at these rates, and is shielded from radiation a good part of the time.

they are frequently clouded by biased technical arguments from one side or the other. You, as a citizen, may have a hard time sorting the real from the imagined, the possible from the probable; the more complete your own understanding of technology and the underlying science, the better you will do in making the proper interpretations of the facts to use in deciding how best to make society move in the directions of your values. This, of course, is what this book is all about.

PROBLEMS

Some practice with numbers

7.1 Calculate the kinetic energy of a 1,000-gram body moving at 100 cm/sec, and express the results in electron volts.

7.2 Calculate the energy of a proton moving at 50% of the speed of light; express in Mev.

7.3 The isotope of sodium, Na^{24}, decays with a half-life of 15 hr. Suppose that leak deposits 1 millicurie into the environment. How much radioactive material will remain 60 hr later?

7.4 Recall that 1 "rad" deposits 100 erg/gram of tissue. Suppose that each gram occupies 1 cubic centimeter. If the energy of the particles is 1 Mev, how many particles must strike the cube of tissue to deposit 500 mrem?

7.5 Suppose a small nuclear accident releases 10^{20} atoms of radioactive atoms, and that a reactor operator somehow takes these onto his person. Suppose they decay by emission of 2-Mev particles with a half-life of 30 min, and that all their energy is absorbed by his tissue. Calculate his radiation exposure in rads for this 30-min interval.

To acquaint you with the nuclear power industry

7.6 Browse through a copy of "Nuclear Energy Conversion," by M. N. El-Wakil, Intext Publishers, New York, 1971. Write a short descriptive article on some interesting aspect of nuclear-reactor power that goes beyond the material presented above.

7.7 Look through some recent copies of *Nuclear News*. Write a short review of the current areas of concern and interest in the nuclear power field based on what you see in this trade publication.

7.8 Write to a manufacturer of nuclear-reactor systems, such as General Electric, Westinghouse, or Gulf General Atomic. Ask for descriptive material on their latest commercial power reactors. Then, write a description of this system in terms that would be understandable to a layman.

7.9 Visit a nuclear power plant. Draw up a flow diagram of the system; find out the capital costs, safety history, etc. Write a descriptive article on this plant that would be suitable for publication in a magazine devoted to informing the public about energy technology.

7.10 Read Fast Breeder Reactors, *Scientific American*, September 1970, and The LMFBR, *Mechanical Engineering*, February 1973. Then, read a more recent article on breeder reactors in a magazine such as *Nuclear News*. Write a short description of breeder-reactor technology, emphasizing the evolution that you see taking place.

Some more comprehensive research problems

7.11 Find out all you can about the "emergency core cooling" provisions for modern reactors. What accidents have occurred in which these systems have been called upon where they failed? What radiation exposure resulted? What is the public impression about these systems? Is the public adequately informed, or has it been misled by the industry or its attackers? Write a paper on your findings.

7.12 Find out all you can about the current status of fusion power research. How much money is spent annually on research related to the development of fusion power? What rate of progress seems to be taking place? What new ideas are exciting those working in this area? What disappointments have these workers had in the past? How do they feel about it? Write a paper on your findings.

7.13 Visit a university engineering or hospital medical laboratory where radioactive materials are used. Find out how they are used, how they are stored, and how theft and spillage are prevented, etc. Write a paper on the health hazards of this operation and the benefits that the operation affords.

7.14 Read the advertisements in *Physics Today*, *Nuclear News*, and other magazines serving the nuclear industry. What ancillary industries have sprung up around the nuclear power business? What new technologies have developed as a result of research in nuclear energy? Write a paper describing the impact of the nuclear energy industry on other segments of the economy.

7.15 Get to know some nuclear engineers, either through their companies or through professional meetings such as local meetings of chapters of the American Nuclear Society. Find out what sorts of things interest them. How do they feel about working on weapons? How do they feel about the safety of nuclear power reactors? How do they feel about the management of the AEC? How do they compare with other people whom you know? Write a paper relating your interpretation of their value systems and the way that they have adjusted in their own particular field.

8

IT HAPPENS
ONLY ON EARTH

In which we see
the power of the world's
greatest engineer

ENERGETICS OF THE ATMOSPHERE

The earth is a very unique place. It is a planet of a particular size which
gives it a particular gravitational strength. It is made up of particular
elements, which gives it a particular chemistry and a particular atmosphere. And it is located at a particular distance from a star with a particular rate of radiant energy release, which serves to drive its atmosphere
in a particular way. This in turn makes particular chemical evolutions
more likely than others, and makes for particular climates, and all this
has led to the particular forms of life that presently inhabit the earth.
In this chapter we shall acquaint you with the energy flows that regulate
the earth's atmosphere, oceans, and life systems. In Chap. 1 we looked

briefly at the overall energetics of the earth. We learned that there is very nearly an exact balance between the income of energy from the sun and the output of energy by radiation to deep space. This delicate balance determines the average temperature of the earth, and small changes in either the input or output would result in changes in the earth's temperature. Man is doing things that can produce these changes, and to understand why, how, and how much, we need to know more about the details of the energy flows within the atmosphere. So, we will begin with a more comprehensive look at what happens to the incoming radiation.

Radiation is produced when atoms in excited states decay to lower-energy states by emission of photons. If the photon carries a lot of energy, it will appear as high-frequency (or short-wavelength) radiation. So, in a very hot body such as a star the atoms are in very energetic, excited states, and thus they emit radiation at very short wavelengths. In contrast, a colder body, such as the earth, will emit radiation at longer wavelengths. The property of the radiation that describes quantitatively the relative amounts of radiation at different wavelengths is called the "spectrum"; the "spectral density" is defined as the amount of radiant energy in a narrow band of wavelengths, expressed on a per-wavelength basis. Figure 8.1 shows the spectral-density distribution with wavelength (the "spectrum") for solar radiation at the point where this energy is incident upon the earth's disk. The solar spectrum is sharply peaked at just below 1 micron. About half the solar energy is in the visible range (0.4 to 0.7 micron). The rest of the solar energy can be felt but not seen, and most of this unseen energy is in the "infrared" range over 0.7 micron. The solar spectrum corresponds to radiation from a body at a temperature of around 6000°K. In contrast, the earth, which has an average radiation temperature of around 285°K, radiates energy at much longer wavelengths. The earth's radiation is almost entirely invisible, with practically all the energy lying in the infrared range. A spectrum for radiation of the same power from a body at 300°K is also shown in Fig. 8.1. There is, of course, a tiny amount of radiant energy in the visible range; otherwise things could not be seen!

The rate of energy incidence on the top of the atmosphere per square meter of surface area varies markedly with latitude, season, and time of day. The radiative power per unit of area (the radiative "flux") from the sun is the same at all points on a plane perpendicular to the radiation (plane *AA* in Fig. 8.2). Along the path of the sun's declination this energy also passes through the same area on the surface (area *B* of Fig. 8.2). However, near the poles this energy is spread out over a large area (*C* in Fig. 8.2), and hence the power per unit of *surface* area is lower near the

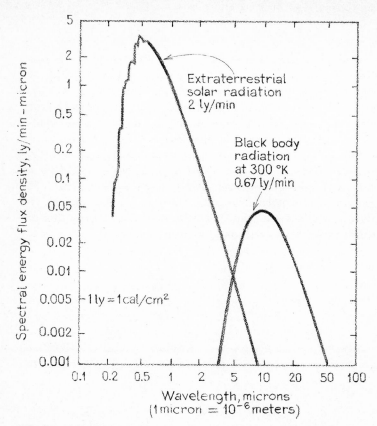

FIG. 8.1 SPECTRA OF SOLAR AND TERRESTRIAL RADIATION.

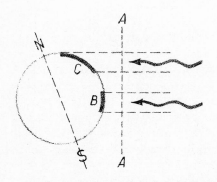

FIG. 8.2 GEOMETRICAL SPREADING.

poles. The length of time each day that the area near the poles is exposed to the sun is quite different from that at the equator. In fact, within the arctic circle the sun never sets in the summer. So, the net result is that each day in June more energy is incident on a square mile of earth at the North Pole than at the equator! Indeed, the world's highest amount of incident energy per day occurs at the South Pole on Dec. 22! Figure 8.3 shows the daily energy-flux input to the top of the atmosphere as a function of latitude and season. These values are in "langleys," a unit of energy flux popular with atmospheric scientists; 1 langley = 1 cal/cm². The solar energy flux (perpendicular to plane *AA* in Fig. 7.2) is about 2,880 langleys/day, which is more than twice the maximum energy flux observed at a fixed point on earth on any given day.

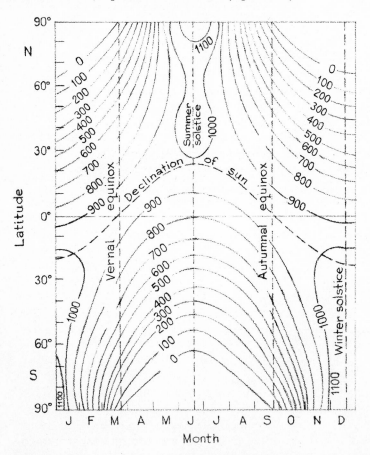

FIG. 8.3 DAILY SOLAR-RADIATION INPUT AT THE TOP OF THE ATMOSPHERE, IN LANGLEYS.

What happens to the solar radiation incident upon the atmosphere? Some of the solar energy is reflected and scattered by the atmosphere and clouds back into space. Some penetrates directly through to the surface as direct-beam solar radiation, and some is scattered by the atmosphere and ultimately reaches the surface as diffuse sky radiation. This scattering is selective in the blue range in a normal cloudless smog-free atmosphere, which is why the sky looks blue (you can guess why it looks brown on a smoggy day).

Absorption plays several key roles in the atmosphere. The various molecules in the atmosphere absorb radiation in their own way; this jumps them to excited states. High in the atmosphere O_2 (oxygen) and O_3 (ozone) molecules absorb some of the incoming radiation, mostly in the ultraviolet portion of the spectrum. This effectively blocks out all radiation at wavelengths shorter than 0.3 micron, and provides the main source of energy to drive the air motions in the upper atmosphere. In addition, it shields biological systems on the surface from damaging "ultraviolet" (short-wave) radiation. Most of the atmosphere absorption is due to dust, water, and carbon dioxide in the air, and occurs throughout the lower levels of the atmosphere. At wavelengths about 0.7 micron the solar beam is strongly depleted through absorption by water and carbon dioxide. This varies from place to place, depending on the local climate. The clear atmosphere absorbs better than do clouds, which in part explains why it can be warm on cloudy days.

Figure 8.4 shows the average distribution (by percentage) of the incident solar energy. Note that only about half the incident energy reaches the earth's surface.

This is only part of the radiation picture in the atmosphere. The earth's surface and atmosphere are warm, and hence both radiate energy. Because of the low temperature this energy is long-wavelength, as discussed previously. Some of this energy passes directly to space; the rest is absorbed by the atmosphere, or scattered within the atmosphere and eventually absorbed by the atmosphere or returned to the surface. The atmosphere itself radiates energy at various rates from various regions. Some of the atmospheric radiation escapes to space, some is reabsorbed elsewhere in the atmosphere, and some is returned to earth. These radiative interactions between the land and ocean surfaces and the atmosphere are very complex, and a proper accounting requires tracing all the bounces that take place. Figure 8.5 shows the average distribution of terrestrial and atmospheric radiation, expressed in percent of the energy incident on the top of the atmosphere.

If you examine the numbers in Figs. 8.4 and 8.5, you will see that the atmosphere exports considerably more radiant energy than it receives.

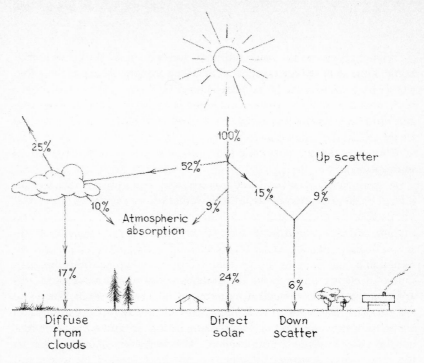

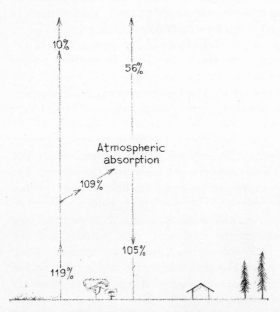

FIG. 8.4 DISTRIBUTION OF THE INCOMING SOLAR RADIATION.

FIG. 8.5 DISTRIBUTION OF TERRESTRIAL AND ATMOSPHERIC LONG-WAVE RADIATION, EXPRESSED IN PERCENTAGES OF THE INCIDENT SOLAR ENERGY.

There most be other mechanisms of energy transport between the surface of the earth and the atmosphere to supply the extra energy. Convective heat transfer is one of these mechanisms; but the primary mechanism is the transfer of energy between surface and atmosphere associated with the removal of water vapor from the surface and the return of water, liquid (rain) and solid (snow).

For the earth as a whole the radiation incident at the top of the atmosphere exactly balances the radiation from the atmosphere. But this is not true locally or seasonally. Toward the poles from 40° latitudes the radiation from the atmosphere is larger than the solar radiation. So, some energy must move from more equatorial latitudes to these polar regions to supply the energy to the atmosphere in these regions. In the region between the 40° latitudes the radiation output is smaller than the solar input, and so some energy must be carried away from these regions, or else stable temperatures would not be obtained. The strong convective circulations in the atmosphere (and oceans) are responsible for these energy transfers; if convective transport of energy did not take place in the atmosphere, the surface would be warming up at the rate of 250°C per day in some places and the atmosphere would be cooling by as much as 1°C per day in others!

The mechanisms by which convective circulations take place in the atmosphere are reasonably well understood. In fact, mechanicians are now able to describe atmospheric motions with mathematical equations, and solve these on large computers. Such calculations today form a primary basis for long-range weather forecasting! The actual flows are very complex; we'll use Fig. 8.6 to help explain what happens. Air near the surface on the summer side of the equator is warmed and tends to rise because it is lighter than the surrounding air mass. This air will

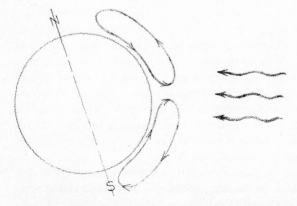

FIG. 8.6   LARGE-SCALE CIRCULATIONS IN THE ATMOSPHERE ARE RESPONSIBLE FOR MAINTAINING LOCAL ENERGY BALANCES.

expand as it rises and eventually fan out to flow toward the poles along the top of the atmosphere. As it cools by radiation to space, it will tend to sink and eventually returns to the surface, and then flow back toward the equator. This cellular circulation pattern is strongest in the winter hemisphere; usually several secondary circulation patterns are formed, depending upon the season. This sort of model was postulated by George Hadley, an English meteorologist, in 1735, and the cellular circulations bear the name "Hadley cells." The general picture has been confirmed by many experiments and theoretical calculations.

Superimposed on this general north-south circulation is a complex pattern of waves called "Rossby" waves. These waves progress from west to east and are responsible for the prevailing motion of the atmosphere in this direction. Waves with length scales of a few thousand kilometers progress around the globe in times of the order of one week. Typically there are about seven waves wrapped around the globe. Since they generally develop as a result of strong horizontal temperature variations, they are most pronounced in winter months. The cyclones and anticyclones that appear in weather patterns are manifestations of these waves, and are very effective in providing the horizontal convective transports of energy that are needed to offset the local radiation imbalance.

How does the energy convected about by atmospheric circulation get off the surface, and how is it deposited in the atmosphere? Water plays a very key role in this process. Solar heating of oceans and lakes causes water to evaporate; water vapor (steam) is then carried about by atmospheric currents. An engineer would think in terms of the enthalpy (flow energy) associated with this stream; meteorologists think instead of the "latent heat" of the water vapor, which is simply the enthalpy change associated with evaporation. In addition, the air is warm at the equator, and warm air has more energy than cold air. Meteorologists call the enthalpy associated with the warm air the "sensible heat" of the atmosphere; you will see various terms like these in your reading of atmospheric energetics, depending on the background of the particular writer. In any event, water vapor is carried from the equatorial regions to the more polar latitudes, where it rains or snows. The condensation or freezing of the water takes energy out of the water; where does this energy go? We hope that you guessed correctly that it is put into the air, and that this supplies a major portion of the energy radiated by the atmosphere in the polar latitudes.

Man's use of energy unquestionably affects the atmospheric circulation, weather, and climate, and the only real question is how much and in what direction? There is a great deal of debate on this question, and the complexity of the processes involved means that simple answers can't be obtained. For example, there is a big question about the effect of carbon dioxide in the upper atmosphere; the amount of CO_2 in the atmosphere has risen at the rate of about 0.2% per year since the early 1940s. Now, CO_2 is an excellent absorber of long-wave radiation, of the type emitted from the earth's surface. Since the CO_2 content is increasing, there should be more atmospheric absorption, and hence the "greenhouse" effect should tend to make the earth get warmer. But, the average temperature of the atmosphere seems to have fallen slightly in this period. Does the extra CO_2 help or hurt? It is not really clear at this point. This is a question of great interest to atmospheric scientists, and of great concern to the public who live and try to survive under the CO_2 roof above.

Local climatological changes are more likely to be of immediate concern. But no one has investigated what would happen to the weather in Arizona if large portions of that state were to be used for collection of solar energy for export to Los Angeles and New York. Without the reradiation characteristic of a hot desert area, might not the atmospheric circulation patterns change to bring energy into the region to supply the energy for radiation from the atmosphere? This would mean more rain in that area, which could be very beneficial but could also reduce the suitability of the area for solar energy collection! Whether this is a fearful fantasy or a real potential problem is not at all certain; but this is the sort of question that a responsible citizen should ask and insist on good answers for before any large-scale effort at solar energy collection is attempted.

ENERGETICS OF THE BIOSPHERE

A tiny fraction of the solar energy that reaches the surface of the earth is absorbed by plants. Through an elaborate and very complex food chain that mother nature devised through a very long research and development activity, this energy maintains all living things. No study of energy can be complete without some discussion of nature's supreme technology.

The principal process in the energetics of life support on earth is "photosynthesis," the process by which plants convert radiant energy to an easily usable chemical form. Photosynthesis is nature's version of "direct energy conversion," in which radiant energy is converted to a convenient form without an intermediate "heat engine." Nature's "heat engine," the atmosphere, of course plays an important role in life support, bringing rain to thirsty plants, giving mobility to drifting seeds, and providing clouds to shield life from the searing sun.

Let's begin by recalling the basics of chemical reactions as discussed in

Chap. 5. Molecules are formed from atoms, and are held together by chemical bonds. These bonds take the form of electrons in the outer shells of the atoms that are "shared" by the atoms in the molecule. Now, any process that upsets these shared electrons can assist in the breakup of the molecule; any process that places electrons in a better position of being shared can assist in the formulation of new molecules. And this is precisely what is involved in photosynthesis.

Photosynthesis is a complicated stepwise process. It begins when light energy is absorbed by chlorophyll molecules in plants. One can think of light as a stream of tiny particles, "photons," that carry kinetic energy and momentum, but not mass. Photons strike chlorophyll molecules and interact with the outer-shell electrons, and this helps the chlorophyll molecules decompose and react with other molecules in the plant to form a variety of other molecules, of which carbohydrates and adenosine triphosphate (ATP for short) are the most important. The net result is an increase in the energy stored within the plant; some of the light energy has been converted to chemical bonding energy, and it is this energy that the plant uses for its life and that animals that eat the plant obtain for their survival.

Plants do not absorb all the incident radiation, and they do not convert all that they absorb to chemical bond energy. Only relatively high-energy photons are able to disturb the chlorophyll molecules sufficiently to initiate photosynthesis; light in the violet end of the visible part of the radiation spectrum and invisible radiation in the ultraviolet range are most effective in promoting photosynthesis. Something of the order of 5% of the solar radiation incident on an average planted area is converted to chemical energy by the plants. However, there is wide variation between plants of different types; some plants evidently are able to convert as much as 15% of the incident radiant energy to chemical bond energy through the photosynthesis process.

The chemical composition of the atmosphere plays a crucial role in the energetics of plant life. Carbon dioxide and water vapor in the atmosphere absorb radiation in the infrared range, and ozone absorbs much of the ultraviolet radiation. Thus, the atmosphere provides a "window" for radiation in the intermediate visible range. It is not surprising that mother nature designed eyes for her creatures that are sensitive in this intermediate range; given a different composition in the atmosphere, different wavelengths of radiation would get through, and thus different eyes would be required and different energetic processes would have to evolve. The blocking of ultraviolet radiation by the atmosphere protects living systems from excessive doses of high-energy radiation, which can produce deathly breaking of molecular bonds. The radiation emitted from the earth's surface is almost totally in the infrared region; if it were not for the opacity of the atmosphere to infrared radiation, this energy would all be lost to space, nights would be very, very cold, and plants would freeze and die. So, life depends upon both the opacity and the transparency of the atmosphere; if the wastes of human activity that are discharged into the atmosphere change its character significantly, life will be unable to exist in its present form. Since man is apt to produce changes much faster than mother nature can alter her technology, it is entirely possible that man's failure to use his technology to maintain the character of his atmosphere could lead to the death of all life on earth.

Animals and plants work together to maintain a very delicate chemical balance in the atmosphere. The process of photosynthesis is very complicated, and involves a number of substeps; but the overall chemical reaction can be represented by

$$6CO_2 + 6H_2O + \text{solar energy} \rightarrow C_6H_{12}O_6 + 6O_2$$

The molecule $C_6H_{12}O_6$ is "glucose," a sugar and a derivative of carbohydrates (CH_2O) . The food chain in animals is initiated by those that feed upon plants, extracting the bonding energy provided by the glucose and other carbohydrates. Another series of very complicated chemical reactions support animal life, which returns carbon dioxide to the atmosphere through processes of respiration, which can be represented by

$$C_6H_{12}O_6 + 6O_2 \rightarrow 6CO_2 + 6H_2O + \text{energy}$$

Thus, animals supply the CO_2 needed by plants for photosynthesis, and plants supply the oxygen needed by animals for respiration. The sun provides energy for plants, the plants provide the energy source for herbivores, and herbivores provide the energy source for carnivores, which in turn provide the energy source for other carnivores, including man. Dead plants and animals provide the food for decomposing organisms, which thereby return the chemical ingredients of life to the soil to be used by plants another day. For all life the ultimate source of energy is the solar energy passed down to earth through the atmosphere's delicately balanced radiation window.

The details of the mechanisms by which individual cells in plants and animals engage in these energy-exchange processes have been the subject of intense research in recent times, and these processes are now rather well understood. The energy-exchange mechanisms of living cells are of two types. "Autotrophic" (self-reliant) cells get their energy directly from sunlight. "Heterotropic" cells require a supply of ready-made fuel of considerable chemical complexity, such as protein and carbohydrate, which themselves are derived from other cells. Heterotrophic cells obtain energy by oxidizing these fuels with oxygen drawn from the atmosphere by respiration, converting some of this energy to forms useful for the biological creation in which they reside.

Chlorophyll molecules are connected spatially with other molecules within autotrophic cells in such a way that electrons jarred free from the chlorophyll molecules by photons are carried away and passed around by other molecules through a circular chain of chemical reactions. The electrons lose their energy bit by bit in these reactions, forming ATP molecules in the process. The ATP molecules store energy in the form of high-energy chemical bonds between oxygen and phosphorus ("phosphate chemical bonds"). The ATP is used in a second series of reactions where carbon dioxide and water are consumed, glucose is formed, and oxygen is released. Thus, the autotrophic cells can be thought of as miniaturized little chemical plants that convert sunlight directly into chemical-bond energy of glucose. Under special laboratory conditions it appears that the efficiency of this conversion process can be as high as 75%!

Heterotrophic cells are tiny energy-conversion systems of a different type. Glucose acquired by the cell and ATP within the cell are reacted chemically through a long series of reactions to create lactic acid and to double the amount of ATP available. This process, called "glycolysis," in essence converts chemical-bonding energy of glucose into the more easily tapped chemical-bond energy of ATP. A second complex series of chemical reactions, the "citric acid cycle," finally oxidizes the products of glycolysis into carbon dioxide and water, and produces additional ATP molecules. Thus, the overall process is one of conversion of the bonding energy of chemical energy input in the form of glucose to chemical energy in the form of ATP. The efficiency of this little chemical-energy-conversion system is of the order of 50%! The complex miniature structure of these two types of energy-conversion systems is truly remarkable, and their high efficiency must make the twentieth century engineer very humble. Advances in electronic technology have produced superminiaturized components, but these are giants in comparison with the remarkable miniaturization of the complex energy-transforming systems that have been developed by mother nature within every living cell; nature is truly the world's most accomplished engineer!

Some very exciting research into the processes of photosynthesis, aimed at replicating it in a man-made device, is now well along. Perhaps within the decade a new energy conversion technology, based on nature's amazing designs, will develop to the point where very high efficiency solar energy conversion is possible and practical; this could revolutionize our thinking about our choices for energy systems!

FROM NATURE TO MAN

Man survives by eating a combination of plants and animals. We have discussed some of the processes by which solar energy is converted to useful chemical energy in plant and animal cells; but what about the efficiency of all this from the point of view of man? Popular agricultural crops convert of the order of 1% of the incident radiant energy to chemical energy; something of the order of 10% of the chemical energy eaten by a person is used by the body, the rest being discharged in chemical wastes and heat. So, a vegetarian diet represents something of the order of a tenth of a percent efficiency of utilization of solar energy. In societies that rely on wild animals for meat, the creatures that come to their end on the dinner table might have fed on other creatures, which have fed on other creatures, and so forth up the food chain to creatures that fed on plants. A food chain that contains a dozen creatures would have an efficiency of the order of 0.1^{12}, or 0.0000000001%! With such a low efficiency a very large area is needed to collect enough solar energy to sustain each person, and starvation will limit the population of any such area. On the other hand, by growing beef, man has greatly shortened the food chain, to the point where his efficiency of extraction of solar energy is of the order of 0.01%. Thus, in spite of the remarkable efficiency that nature has given us at the cellular level, we still do not do a very competent job in our overall use of the sun's energy.

An average person consumes of the order of 70 watts just to maintain minimum body functions. Activity increases the energy-expenditure rate to as much as 400 watts, i.e., the power used by four typical desk study lamps. The average normal adult in a modern society needs an input of about 3,000 kcal/day to function properly. We can combine this information with the efficiency estimates to calculate the maximum population densities that can be sustained with current agricultural technology. Let's take an area where the average incident solar radiation is about 300 langleys/day (do you remember that 1 langley is 1 cal/cm^2?). At 100% efficiency of solar energy conversion, it would take 10^4 cm = 1 m^2, or about a square yard, to collect enough solar energy to maintain one person. However, in a hunting society with an overall efficiency of 0.1^{12}, it will take 10^{12} square meters, or about half a million square miles, to produce enough food to support a single person! In contrast, a society based on beef production will have an efficiency of about 0.01%, and each square mile of pasture will be able to support about 250 persons; in other words, each person would require about 1 acre of land for beef production. There are some 5 billion acres of grazing land in use today, and it has been estimated that 8 billion acres are available. This suggests that the world should try to limit its population to around 8 billion people, slightly more than twice the present population. The alternatives are to develop more efficient food chains, or to survive on less, but even with some reasonable adjustments it is not likely that the earth can support more than a dozen billion people. Without population control, the world will suffer famines that have their fundamental roots in the inability of the combined technologies of man and nature to convert a sufficient fraction of the incident solar energy to forms suitable for the sustenance of life. Mother nature, the grand engineer, may someday develop new technologies to deal with these conversions more efficiently, but her development times are measured in eons rather than years. It will fall upon man's technologies to improve the efficiency of energy utilization for food production, and to limit population growth, in order to develop a satisfactory steady-state existence for life on earth.

Man's technology is presently depleting the energy reserves of the earth. Fusion offers the only long-term hope as a reserve energy source; but the use of any reserve energy leads ultimately to the deposition of waste heat into the atmosphere, and this has already led to local weather complications. Man should perhaps take a clue from mother nature. She did not develop creatures that use the energy reserves of the earth; instead she developed complex systems for direct and indirect use of the income energy from the sun. Solar energy has been neglected by man as too difficult to use, too costly, too impractical. As long as reserves are available, and as long as the use of these energy reserves does not significantly alter our atmosphere or biosphere, these are rational arguments. But what should man do when these arguments no longer have strength?

Man can capture solar energy with his own technology, using mirrors to collect the energy, pipes and pumps to circulate fluids through the focal points of the mirrors to extract the solar energy as heat, and sophisticated vapor power cycles to produce electrical energy. Estimates reveal that large areas of reflectors will be required; who is going to keep these mirrors clean? An alternative that deserves considerable study involves man's use of nature's technology to use solar energy. There are efficient chemical processes for producing alcohol and methane from glucose, and both alcohol and methane are easily stored and transported. Let's see how much alcohol and methane we would have to grow to supply our energy needs using agriculture. The world energy use in 1970 was of the order of 2×10^{19} calories. The solar energy incident on the approximately 4 billion acres of land presently under cultivation was of the order of 5×10^{22} calories. If of the order of 4% of this land were put into agricultural·production of crops that yield alcohol and methane at about 1% efficiency, all the energy used by all the world in 1970 could be provided. And there would be no mirrors to wash! In the United States we pay farmers *not* to produce crops for food; we could instead pay them to produce crops for energy!

The agricultural production of solar energy is a concept that should receive considerable study and attention in the years ahead. It represents a way to combine man's best technologies with the best technologies of nature, and may be the only long-term way to safely, reliably, and cheaply bring the full benefits of abundant and useful energy from nature to man.

PROBLEMS

You are now on your own

8.1 Read The Energy Cycle of the Earth, and Human Energy Production as a Process in the Biosphere, both in *Scientific American*, September 1970, and Thermal Pollution and Aquatic Life, *Scientific American*, March 1969. Then, write a short article for your local newspaper outlining the major problems that man's use of energy has caused or may cause.

8.2 Read The Flow of Energy in a Hunting Society, The Flow of Energy in an Agricultural Society, and The Flow of Energy in an Industrial Society, all in *Scientific American*, September 1971. Then, write a short article for *Reader's Digest* that summarizes the differences in the energy flows in these societies and discusses the implications for the future development of the world.

8.3 Read How Animals Run, *Scientific American*, May 1960, The Energetics of Bird Flight, *Scientific American*, May 1969, The Flow of Energy in the Biosphere, *Scientific American*, September 1971, and How Cells Transform Energy, *Scientific American*, September 1961. Then, write a short article on Nature's grand technology.

8.4 Survey *Scientific American* issues published since January 1972. Review the articles that deal with nature's energy technology, and those that deal with man's energy technology. Discuss and compare the systems that you have learned about through this review.

8.5 Prepare a short talk on energy as it relates to man's problems suitable for presentation to an eighth grade class. Decide for yourself what material is important to cover, what reading you can recommend to the students, and what reading you can suggest to their teacher to help increase his or her effectiveness in dealing with energy as a class study subject. Your local school district will probably be delighted to accept your offer of a lecture on this subject.

8.6 Find a newspaper article discussing a particular energy problem of interest. Does the article appear to make sense scientifically and technically? What aspects of the problem that were neglected by the article do you feel are important? Write a short letter to the editor detailing your contribution.

8.7 Survey your magazines for advertisements that relate to energy. Critique each in terms of its scientific accuracy, technical sensibility, operative system, and probable impact upon its readers. Write a short paper summarizing your findings, and send copies to the publishers of each magazine.

8.8 Read your electricity and gas meters, and live a normal week. Read them again, and live a week in which you take special care to conserve energy. Read the meters at the end of the week, and compare your energy expenditures for the two weeks. Then, set a goal of an additional 50% saving of energy, and plot your daily expenditures of energy to see how you are doing during the third week. Force yourself to meet your energy-conservation goal. Write a short article on your experiences for submission to your local newspaper, and send copies to your local utility companies.

8.9 Eat normally for one week, keeping an accurate calorie count. Then, maintain a diet of only 2,000 kcal/day for a few days (give it up when you begin to get adverse health effects, and don't start without your doctor's approval!). Write a short article indicating how this low-level diet affected your performance; include comparisons with diets available to persons in less fortunate circumstances.

8.10 Read "The Limits of Growth," Universe Books, 1972. What does this book say about the outlook for mankind? What role does energy play in the analysis? Read some published reviews of the book, and then draw your own conclusions about the predictions made by the authors. Think about the role of technology in solving these problems, and what you would do if asked to vote upon national policies of the type recommended by the authors for the development of a more stable world.

Appendix

UNIT SYSTEMS AND CONVERSION FACTORS

Table A.1
Mechanical Unit Systems

	Mks	Cgs	Absolute engineering	Engineering*
Primary quantities and their units:				
Length	Meter, m	Centimeter, cm	Foot, ft	Foot, ft
Mass	Kilogram, kg	Gram, g	...	Pound mass, lbm
Time	Second, sec	Second, sec	Second, sec	Second, sec
Force	...	...	Pound force, lbf	Pound force, lbf
Newton's second law, $F = \dfrac{1}{g_c}\, ma$	$g_c \equiv 1$ (selected)	$g_c \equiv 1$ (selected)	$g_c \equiv 1$ (selected)	$g_c = 32.17$ ft-lbm/lbf-sec^2 (experimental)
Secondary quantities and their units:				
Force	kg-m/sec^2	g-cm/sec^2	lbf-sec^2/ft	
Mass	...	...	...	
Energy, $W = FX$	kg-m^2/sec^2	g-cm^2/sec^2	ft-lbf	ft-lbf
Power, $W = W/t$	kg-m^2/sec^3	g-cm^2/sec^3	ft-lbf/sec	ft-lbf/sec
Aliases:				
Force	1 newton $\equiv$ 1 kg-m/sec^2	1 dyne $\equiv$ 1 g-cm/sec^2		
Mass	...	...	1 slug $\equiv$ 1 lbf-sec^2/ft	
Energy	1 joule $\equiv$ 1 kg-m^2/sec^2 = 1 newton-m	1 erg $\equiv$ 1 g-cm^2/sec^2 = 1 dyne-cm		
Power	1 watt $\equiv$ 1 kg-m^2/sec^3 = 1 joule/sec			

* A body having a weight of 1 lbf on the surface of the earth will have a mass of approximately 1 lbm.

Table A.2

Electromagnetic-Unit System

Primary quantities and their units:

Length	Meter, m
Mass	Kilogram, kg
Time	Second, sec
Charge	Coulomb, coul

Coulomb's law

$$k_C = \frac{1}{4\pi\varepsilon_0}$$

$$F_{12} = k_C \frac{Q_1 Q_2}{r_{12}{}^2}$$

$$\varepsilon_0 = 8.854 \times 10^{-12} \text{ coul}^2\text{-sec}^2\text{-kg}^{-1}\text{-m}^{-3}$$

Secondary quantities and their units:

Current, $I = Q/t$	coul-sec^{-1}
Electric field E	kg-m-coul^{-1}-sec^{-2}
$\quad F = QE$	
Electrical potential V	kg-m^2-coul^{-1}-sec^{-2}
$\quad V = EX$	coul^2-sec-kg^{-1}-m^{-3}
Resistance, $R = V/I$	kg-m^2-coul^{-2}-sec^{-1}
Capacitance	coul^2-sec^2-kg^{-1}-m^{-2}
Inductance	kg-m^2-coul^{-2}

Aliases:

Current	1 amp $\equiv$ 1 coul-sec^{-1}
Electric potential	1 volt $\equiv$ 1 kg-m^2-coul^{-1}-sec^{-2}
Resistance	1 ohm $\equiv$ 1 kg-m^2-coul^{-2}-sec^{-1}
Capacitance	1 farad $\equiv$ 1 coul^2-sec^2-kg^{-1}-m^{-2}
Inductance	1 henry $\equiv$ 1 kg-m^2-coul^{-2}

Table A.3
Selected Dimensional Equivalents

Length	1 m = 3.280 ft = 39.37 in
	1 cm $\equiv 10^{-2}$ m = 0.394 in = 0.0328 ft
	1 mm $\equiv 10^{-3}$ m
	1 micron (μ) $\equiv 10^{-6}$ m
	1 angstrom (Å) $\equiv 10^{-10}$ m
Time	1 hr $\equiv$ 3,600 sec = 60 min
	1 millisec $\equiv 10^{-3}$ sec
	1 microsec (μsec) $\equiv 10^{-6}$ sec
	1 nanosec (nsec) = 10^{-9} sec
Mass	1 kg $\equiv$ 1,000 g = 2.2046 lbm = 6.8521×10^{-2} slugs
	1 slug $\equiv$ 1 lbf-sec^2/ft = 32.174 lbm
Force	1 newton $\equiv$ 1 kg-m/sec^2
	1 dyne $\equiv$ 1 g-cm/sec^2
	1 lbf = 4.448×10^5 dynes = 4.448 newtons
Energy	1 joule $\equiv$ 1 kg-m^2/sec^2
	1 Btu $\equiv$ 778.16 ft-lbf = 1.055×10^{10} ergs = 252 cal
	1 cal $\equiv$ 4.186 joules = 0.00397 Btu
	1 kcal $\equiv$ 4,186 joules = 1,000 cal
	1 erg $\equiv$ 1 g-cm^2/sec^2
	1 ev $\equiv 1.602 \times 10^{-19}$ joules
Power	1 watt $\equiv$ 1 kg-m^2/sec^3 = 1 joule/sec
	1 hp $\equiv$ 550 ft-lbf/sec = 33,000 ft-lbf/min
	1 hp = 2,545 Btu/hr = 746 watts
	1 kw $\equiv$ 1,000 watts = 3,413 Btu/hr = 1.3405 hp
Pressure	1 atm $\equiv$ 14.696 lbf/in^2
	1 mm Hg = 0.01934 lbf/in^2
	1 dyne/cm^2 = 145.04×10^{-7} lbf/in^2
	1 bar = 14.504 lbf/in^2 $\equiv 10^6$ dynes/cm^2
	1 micron (μ) $\equiv 10^{-6}$ m Hg = 10^{-3} mm Hg
Volume	1 gal $\equiv$ 0.13368 ft^3
	1 liter $\equiv$ 1000.028 cm^3

READINGS

For a quaint peek at energy sciences:

"Order and Chaos," by S. W. Angrist and L. G. Hepler, Basic Books, Inc., New York.

For more background on energy technology:

"Energy and the Future," by A. L. Hammond et al., American Association for the Advancement of Science, Washington, 1973.

For more on energy resources:

"Energy in the United States: Sources, Uses and Policy Issues," by H. H. Landsberg and S. H. Shurr, Random House, New York, 1968.

"Resources and Man," by the Committee on Resources and Man, W. H. Freeman and Company, San Francisco, 1969.

For the environmentalists view:

"Energy," by J. Holdren and P. Herrera, Sierra Club, San Francisco, 1971.

For more on the earth's energy systems:

"Environmental Geoscience: Interaction between Natural Systems and Man," by A. N. Strahler and A. H. Strahler, Hamilton, Santa Barbara, 1973.
"Energy Exchange in the Biosphere," by D. M. Gates, Harper and Row, New York, 1962.

For an analytical engineering view of the social science of energy:

"Environment, Power and Society," by H. T. Odum, Wiley-Interscience, New York, 1971.

INDEX